Routledge Revivals

Easy Lessons in Einstein

First published in 1922, this book represents the first attempt to popularise the more accessible aspects of Albert Einstein's general theory of relativity. Eschewing the mathematical components that put the theory beyond many people's grasp, the author employs metaphorical examples and thought experiments to convey the fundamental ideas and assertions of one of physics' most famous principles — which remains the accepted description of gravitation more than a century after its first publication. This book will of interest to students of physics as an introductory basis to aid further study.

T0188361

Easy Lessons in Einstein

A Discussion of the More Intelligible
Features of the Theory of Relativity

Edwin E. Slosson

Routledge
Taylor & Francis Group

First published in 1922
by George Routledge and Sons

This edition first published in 2017 by Routledge
2 Park Square, Milton Park, Abingdon, Oxon, OX14 4RN
and by Routledge
711 Third Avenue, New York, NY 10017

Routledge is an imprint of the Taylor & Francis Group, an informa business

© 1922 Edwin E. Slosson

Publisher's Note
The publisher has gone to great lengths to ensure the quality of this
reprint but points out that some imperfections in the original copies may
be apparent.

Disclaimer
The publisher has made every effort to trace copyright holders and
welcomes correspondence from those they have been unable to contact.

A Library of Congress record exists under LC control number: 20008295

ISBN 13: 978-1-138-28998-7 (hbk)
ISBN 13: 978-1-315-26664-0 (ebk)
ISBN 13: 978-1-138-29007-5 (pbk)

DR. ALBERT EINSTEIN IN HIS STUDY

[front

EASY LESSONS IN EINSTEIN

A DISCUSSION OF THE MORE
INTELLIGIBLE FEATURES OF
THE THEORY OF RELATIVITY

BY

EDWIN E. SLOSSON, M.S., Ph.D.

Author of Creative Chemistry, Major Prophets of To-day, Six Major Prophets, etc.

*With an Article by Albert Einstein
and a Bibliography*

ILLUSTRATED

NEW YORK : HARCOURT, BRACE AND COMPANY

1922

Deepest of all illusory Appearances, for hiding Wonder, as for many other ends, are your two grand fundamental world enveloping Appearances, Space and Time. —CARLYLE.

Henceforth Space in itself and Time in itself sink into mere shadows and only a kind of union of the two can be maintained as self-existent.—MINKOWSKI.

A PREFATORIAL DIALOGUE

(The Purpose of which is to Prevent the Prospective
Reader from buying the Book under False Pretences)

SCENE: A street car in uniform movement of translation in any direction.

TIME: The present.

The Reader (looking over the top of a morning
paper): Here's something queer—a
whole page taken with a new discovery
in physics—"Eclipse Observations Confirm Einstein's Theory of Relativity."
Anything about it in your paper?

The Author: Yes. Here's a cartoon on it by Mc-
Cutcheon.

The Reader: Must be something to it then. Mc-
Cutcheon always knows what's news.
(Reads on with audible fragments)
"Most sensational discovery in the his-
tory of science"—"Greatest achieve-
ment of the human intellect"—"Upsets

Galileo, Newton, and Euclid "—" Revolution in philosophy and theology." It looks as though I ought to know something about this, doesn't it ?

The Author : I think you will have to sometime. And you might as well do it now and get it over with.

The Reader : (running down the column and hitting the high spots) : " Parallel lines meet " —" a man moving with the speed of light never grows old "—" gravitation due to a warp in space "—" length of a measuring stick depends upon direction of its motion "—" mass is latent energy "—" time as a fourth dimension "—why, the man is crazy, isn't he ?

The Author : Well, definitions of insanity are so uncertain that it is not safe to say who is crazy.   But it seems there's method in his madness—otherwise how could he have hit upon the exact extent of the sun's attraction on light ?

The Reader : (Picks up his paper and reads aloud with concentrated attention) : " Postulate I.   Every law of nature which holds good with respect to a coordinate system K must also hold good for any other system K¹, provided that K and K¹ are in uniform movement of translation." Say, do you know anything about this business?

The Author: Well, yes, a little. I have followed the controversy—at a safe distance—for a number of years.

The Reader: Can you tell me in plain language what it is all about?

The Author: Yes. Just that. I can tell you what it is *about*, though I can't tell you what it *is*. Einstein says that there are only twelve men in the world capable of understanding his latest paper.

The Reader: Are you one of the twelve?

The Author: No, nor the thirteenth. But without plunging into the mathematics of it, we might talk over some of the interestiug aspects of the theory of relativity and in the end I could put you on track of the twelve so you could read up on the subject if you liked.

The Reader: All right. That's fair. This is a slow car anyhow. Go ahead.

The Author: (See following pages)—

EASY LESSONS IN EINSTEIN

" A warp in nature has been found
No line is straight, no circle round;
For Isaac Newton had unsound
Ideas of gravitation."

WHY is it that our newspapers are sending out their reporters to interview astronomers as well as actresses and devoting pages to speculations on the nature of space and time as well as on the state of the market? It is—to get at the bottom of it—merely because a few photographs taken during the eclipse of the sun on May 29, 1919, by two telescopes, one at Sobral in northern Brazil and the other on the island of Principe off the west coast of Africa, showed an abnormal shift of less than one-324,000th of a right angle in the position of the stars. When these photograph films were laid over films taken before the eclipse it was found that the star-images about the darkened disk of the sun did not exactly coincide with the images when the sun was not in their midst. Measured with a micrometer the displacement of the stars from their ordinary positions was found to be 1.60 seconds of arc

A

on the African plates and 1.98 seconds on the Brazilian plates. Average these two observations and you get 1.79. This is extremely close to the 1.73 predicted by Professor Einstein of Berlin and twice as large as the deflection calculated according to Newton's law of gravitation which would be .87 of a second.

When the announcement of this result was made at the meeting of the Royal Society of London on November 6 all eyes were turned toward Sir Oliver Lodge, for last February he had been rash enough to express the hope, if not the prediction, that the results of the eclipse expedition would support Newton rather than Einstein. But instead of taking part in the discussion Sir Oliver got up and walked out. It was suspected that he had "gone off mad," as we Americans would put it, because the starlight would not follow his preferred path. But he put a stop to any such rumours by a letter to *The Times* in which he explains that his departure was not due to any dissatisfaction with the universe but to the necessity of catching the six o'clock train. He frankly acknowledges that " the eclipse result is a great victory for Einstein; the quantitive .87 present is too close to allow much room for doubt," but adds, " a caution against a strengthening of great and complicated generalizations concerning space and time on the strength of this splendid result : I trust that it may be accounted for, with reasonable simplicity in terms of the ether of space."

This caution is wise, but we cannot hold cur breath till 1922, when the next eclipse comes, to see if these observations are verified and we may in the meantime consider some of the implications of Einstein's theory of relativity.

Sir Joseph Thomson, President of the Royal Society, in making the momentous announcement in the session of the Society, said :

If his theory is right, it makes us take an entirely new view of gravitation. If it is sustained that Einstein's reasoning holds good—and it has sustained two very severe tests in connection with the perihelion of Mercury and the present eclipse—then it is the result of one of the highest achievements of human thought. The weak point in the theory is the great difficulty in expressing it. It would seem that no one can understand the new law of gravitation without a thorough knowledge of the theory of invariants and of the calculus of variations.

What is this theory of relativity and why is it so important ? The mathematics of it are too much for most of us, but we can get some notion of it by a familiar illustration.

Suppose you wake up some morning in a Pullman berth and look out of the window to see where you are. You find your view blocked by a passing train on the next track. Now if you do not feel any jar of your car, and cannot catch sight of the landscape beyond the other train you cannot tell whether (1) your train is moving forward and the other train is standing still, or (2) your train is standing still and the other train is moving backward, or (3) whether both trains are moving in opposite directions, or (4) whether both trains are moving in the same direction, but your train faster. It is obvious that the trains are getting past one another. You can measure their speed of parting as accurately as you please. But all you can perceive is the relative motion of the two trains. You begin to wonder whether there is any such thing as absolute motion; whether there is any real difference between rest and motion. Is there any possible way of telling whether your train is in motion or not if all you can see out of the window is some object that itself be moving? Suppose the windows were all curtained, how could you find out whether you were moving forward or backward or standing still?

You discuss this curious question with your fellow passengers at the breakfast table and one of them makes the brilliant suggestion that it might be possible to determine the absolute motion of the car by reference to the air. If the car is moving forward the air would stream from front to rear and the reverse if it were moving backward. "Suppose," says the ingenious experimentalist, " that you stand at one end of the car and I at the other. We will shout at each other alternately and time the passage of the sound with our stop watches. Since sound is carried by air waves it will take longer for the shout to go against the air current than with it, and from that measurement it might be possible for us not only to determine which way the car is moving but how to calculate how fast it travels, assuming, of course, that there is no wind blowing." That strikes you as a crucial experiment, but you point out one possible difficulty, that the doors at the ends of the car may be closed and the air inside is being carried along with the car, so the difference would be observable in the speed of the sound even though the car were moving. "All right," replies your scientific friend, " we will make a preliminary test to see if the enclosed air is carried along with the car, and if we find that it is not then we will try the second experiment with the sound signals to see which way the air current is moving. These two experiments must settle it, for either the air is moving with the car or it is moving through the car. Can you conceive of any other possibility than these two?" No, you cannot, so you proceed to try the two experiments. First you visit both ends of the car and find both doors open; the air then is not being carried along with the car. You turn then with confidence to the second experiment and you find, of course, that there is a difference in the speed of sound whether it moves with the air drift or against it.

There might, I admit, be practical difficulties in the way of carrying out such a delicate experiment on a moving train, but we need not bother with them, for probably the current of air through the car would be so strong as to blow your hat out of the back door and that would settle the question to your satisfaction—or at least it would settle the question in the affirmative.

But imagine your amazement if this second experiment should give negative results like the first one; if you could detect no difference in time whether the sound was sent forward or back or across the car. You would then have proved by experiment (1) that the air did not move with the car and (2) that the air did not move through the car. You might suppose from this that your car is at rest, but suppose the people on the other train passing yours tried the same experiments and got the same result, namely, that they, too, were at rest as regards the air. You would then be in a quandary, for your two indisputable experiments had apparently given contradictory results. You might get out of it by saying that there was no air, but if not what carried the sound waves—and the hat?

CONTRADICTORY EXPERIMENTS

Now this is the quandary in which physicists have been for the last thirty-three years. Is there any way of discovering absolute motion among the heavenly bodies? We can observe and measure with great accuracy their relative motion. The sun is seen to pass across the sky from east to west and man at first assumed that the earth was still and the sun went around it. This is the natural and instinctive assumption, or when you first glance out of your Pullman window you get the impression that the other train is the moving one. But for the last three hundred years it has been the fashion to assume the earth was moving and not the sun. That assumption has the advantage of simplifying the calculations of the astronomers, though I never could see why we should have to give up our simple notions of sunrise and sunset to save them a little trouble figuring.

The earth moves—if it does move—so quietly and silently that we feel no jar or engine-beat to tell us of its motion. If the earth were perpetually shrouded by clouds could we find out its motion through space or even its revolution? And do we actually get any proof on this point from observation of the heavenly bodies? We see them moving about relatively to each other and we can represent their movements most easily by supposing that the moon goes around the earth and that the earth and the rest of the planets go around the sun. But is this whole solar system in motion? So it seems when we compare it with the stars. But who knows if the solar system and all the visible stars are not altogether moving off through space at the rate of a mile or a billion miles a second? How can we tell unless we have something that is still and fixed to measure the motion by?

It seemed until recently that we had such a fixture, the ether. We know of the sun and the stars only from the light that comes from them to us. Light, as we can prove by simple experiments, consists of wave motion. Now, can you have wave motion without something to wave? Sound waves are conveyed by air but there is no air between the earth and the sun. So as nothing could be found to fill this empty space scientists had to invent something to satisfy their sense of the fitness of things. The ether was the product of their excogitations. It was a British invention, devised in the Royal Institution, whence have come so many useful theories and discoveries.

The ether, as Salisbury said, is simply the nominative of the verb " to undulate." It was conceived of as a sort of transparent jelly filling all space, more rigid than any solid, more frictionless than any fluid, more easily penetrated than any gas. It must be more elastic than steel and yet so rarefied that ordinary matter passes through it without the slightest effort. The ether is supposed to slip between the particles of the rushing earth as the wind blows through the branches of a tree.

For many years after its invention the ether had nothing to do except to carry light about from one place to another. But when the electro-magnetic waves of the wireless telegraph were produced something was needed also to carry them and this new task was laid upon the shoulders of the uncomplaining ether. When Röntgen discovered the X-rays, whose waves are 10,000 times shorter than the shortest light waves, these were turned over to

the ether to run. In fact, it got so that whenever a
physicist found any action that he could not explain
by ordinary matter he said : " Let the ether do it,"
and   that   hypothetical   substance   apparently
answered every purpose until it came to this question of relative motion.

Now whatever we may think about the ether it
would seem that if there is any such thing filling all
" empty " space we might use it for measuring the
motion of the earth through it as we did the air current in the car.   If the earth is really revolving
around the sun the ether must be whizzing through
its pores at the rate of about nineteen miles a second.

But wait—there is the possibility that the earth
carries along with it in its flight through space a sort
of atmosphere of ether as it does of air.   We must
first get rid of this possibility by a preliminary experiment to see if a swiftly moving mass of matter
does catch up and carry along with it a little of the
ether. This would cause a sort of an eddy or disturbance in the ether in the neighbourhood of the moving
mass as a boat disturbs the water.   For instance, a
ray of light passing close to a rapidly revolving wheel
would be a little deflected and show a distorted image.
Sir Oliver Lodge tried this experiment and got
negative results. That is, moving matter does not

disturb or carry with it the ether.   Consequently, it would seem, we are left to the only other logical alternative, that the ether drifts through matter and we should expect to detect this drift by measuring the speed of light in the direction of the earth's motion.   It ought to take longer for light to travel from one point to another if the earth meantime is moving away from the first point and it ought to take less time if the earth is moving toward it. Well, Michelson and Morley tried this experiment—and also got negative results! It did not make any difference whether the ray of light was sent in the direction of the earth's movement or the reverse or across the line, it travelled invariably at the same speed, 186,000 miles a second. Here then were two unquestionable experiments apparently contradicting each other. One proved that the ether did not travel with the earth. The other proved that the ether did not stand still while the earth travelled through it.

Now, when we get contradictory answers to the questions we put to nature we must assume—unless Nature is nonsensical—that we are asking nonsensical questions. If in the trial of a pickpocket one witness swears that the thief did not run up the street and another witness that he did not run down the street the lawyer does not necessarily say that one of them must be a liar.   He meditates a moment and then it occurs to him that possibly the pickpocket did not move or that perhaps he disappeared into the third dimension by climbing a fire-escape or dropping into a coal hole.

So with our other quandary.   If the ether does not move and does not stand still perhaps there isn't any ether or perhaps there is a fourth dimension.   These are two conceivable ways out of the dilemma though they are not easy to accept, either of them. If there is no ether what carries the light waves ? If there is a fourth dimension in what direction does it lie ? But it is no harder to believe in or conceive of a fourth dimension than it is the ether, and if the physicist finds that he needs it in his business he will have to have it.   Einstein says that he needs a fourth dimension for his formulas.

THE CONUNDRUM OF THE AGES

For twenty-four-hundred years philosophic thought has been concerned with the problem of the relation of space and time.   Drop into any of the scientific societies of to-day and you will find them discussing whether space is finite or infinite, whether there is any difference between rest and motion, whether length is absolute or relative, whether time and space have real existence, which are the very questions discussed by Pythagoras and Zeno in the Greek cities of Asia Minor. Now the time spent in these speculations has not been wasted, although it has led to no definite conclusion, for out of it have grown our mathematics and physics. The Wandering Jew, who is the only mortal having the privilege of attending the schools of the Eleatics and those of the present day, would observe one difference, that modern scientists try to put their theories to the test of experiment wherever possible, while the ancient were content with thinking them out.

Of all the guesses that have been given to this riddle of the universe none has been more bold and revolutionary than that contained in a paper of four or five pages contributed in 1905 to the *Annalen der Physik* by Albert Einstein. The controversy it precipitated has not altogether been confined to the realm of pure reason, for scientists are but human and as such are not entirely uninfluenced by patriotic prejudice.

In this brief paper he proposed a new theory of the universe based upon two postulates. The first was the principle of relativity; that all *motion is relative*. This means, for instance, that we would never know the motion of a smoothly moving train if the windows were darkened and that we could never discover the forward movement of the earth if we could not see the heavenly bodies.

Einstein's second postulate was that the *velocity of light is independent of the motion of the source.* This is a hard one for our reason to swallow, for it means that nothing can travel faster than light, 186,000 miles a second, and that you cannot make light travel faster than that by giving it a swift send-off.   It is the same as saying that if a man standing on the cow-catcher of an engine threw a ball forward, it would not make any difference with the velocity of the ball whether the train was running at full speed forward or backward or standing still. But the experiments of the American physicists, Michelson and Morley, who measured the speed of light and found it the same whether the earth was moving toward the source of the ray or away from it, or at right angles to its direction, confirm Einstein's second assumption.

If we accept Einstein's two primary postulates and his later "Principle of Equivalence" his theory clears up this ether-drift difficulty as well as various other riddles of the universe. It explains the shifting of the orbit of Mercury that Newton's theory could never account for.   It foretold the direction of light by the sun's gravitation that the observations on the eclipse of last May confirmed. A third test, the shifting of the lines of the solar spectrum toward the red end in a gravitational field, has not been met. Such technical points concern only physicists and astronomers but Einstein's relativity theory, which two out of the three experiments support, carries with it certain speculations as to time and space that are upsetting to current conceptions.

PARADOXES OF RELATIVITY

All three of Newton's laws of motion are now questioned and the world is called upon to unlearn the lesson which Euclid taught it that parallel lines never meet. According to Einstein they may meet. According to Newton the action of gravitation is instantaneous throughout all space.   According to Einstein no action can exceed the velocity of light. If the theory of relativity is right there can be no such thing as absolute time or way of finding whether clocks in different places are synchronous. Our yardsticks may vary according to how we hold them and the weight of a body may depend upon its velocity. The shortest distance between two points may not be a straight line. These are a few of the startling implications of Einstein's theory of relativity. If he had put it forward as a mere metaphysical fancy, as a possible but unverifiable hypothesis, it would have aroused mere idle curiosity. But he deduced from it mathematical laws governing physical phenomena which could be put to the test of experiment. They have been tested in these two crucial cases and prove to be true.

In the preceding pages we have discussed the question of relativity of motion and seen how impossible it is to tell, for instance, whether a train or a ship you are on is moving or not unless you can compare it with something that you are "sure" is stationary. But what are you sure is stationary? Nothing on earth, surely, for the earth compared with the "fixed" stars is spinning around at the rate of about a thousand miles an hour and rushing around the sun at the rate of nearly 70,000 miles an hour. But are we sure the stars are fixed since we have nothing else to compare them with? You may remember Herbert Spencer's illustration of the sea captain who was walking west on the deck of a ship sailing east at the same rate. Is he moving or not? If you are in the same boat, you say he is. If you are on shore when the ship is passing you say he is standing still and "marking time." It all depends on the point of view.

Now you may readily admit that all motion is relative, not absolute, and yet you may balk at the idea that space and time are also relative, not absolute. But motion is merely simultaneous change of position in space and time, and why should we feel so certain about space and time when we have never seen either?

You may say, for instance, that you are sure your desk is *so* long. But if I ask you *how* long you have to say as long as something else. You may say it is a yard long. But how long is a yard? It is as long as some tape or stick marked " one yard," and this in turn has been taken from some other yardstick, until you get back to the brass rod in London that is just as long as the distance from the tip of the nose of King Henry **I** to the end of his royal thumb. But such a standard of absolute measurement is unsatisfactory to everyone except an absolute monarchist. But apart fom the difficulty of the present inaccessibility of King Henry's nose and thumb, can we be confident that our yardstick keeps the same length while we are measuring with it? We must admit indeed that it is longer on a summer day than on a winter day, but can we be sure that it does not alter in length when we hold it upright or lay it horizontally? Or, rather, could we tell if it did change in length as it is changed in direction?

B

ARE YOU SURE OF YOUR SHAPE?

If you have ever been in any of those funny places at the amusement parks you will have noticed the convex mirrors there and how ridiculous they make other people look.   If you cannot afford the nickel necessary for the study of optics in such an establishment you can contemplate your reflection in the side of a shiny cup or can.   In a plane mirror you see a man who looks as you suppose yourself to be except that somehow you seem to have become left-handed. But when you look into a convex cylindrical mirror set

upright you see a man thinner than you "really are."
Look into the same mirror set horizontal and you
see a man shorter than you " really are." You grin
at the sight of such queer-looking creatures, but you
notice that they are equally amused at your shape.
Now how are you going to prove to the men in the
curved glasses that they are mere caricatures and that
you are not really built on the plan of either of these
images? You naturally resort to measurement, as a
scientist should. You cannot get into the mirror
world to measure the tall man who pretends to represent you, but you can explain to him in the sign
language what you want him to do and he instantly
complies. You stand up a measuring rod at your
side and show him that you are exactly 72 inches
tall. He also sets up a rod and that also reads 72
inches. Never mind, let him use any kind of
measure he likes, you will catch him when it comes
to measurement of width with the same stick. You
hold your rule across your shoulders and it reads 18
inches, that is, one-fourth your height. But he also
measures his width with his rule and makes it just
the same, 18 inches, although as you see him he

THE MEASURE OF A MAN

When the man in the middle looks at himself in a curved mirror he sees what he regards as a distorted image. The image on the right is thinner and seems taller because it is reflected from a cylindrical surface set upright. The image on the left is shorter and seems broader because it is reflected from a cylindrical surface set horizontally. But if the man and his image are measured by scales in the real world and the mirror world they come out the same. So, too, it would be impossible for us to find out if everything in the world were expanded or contracted in all directions. In other words, all measurements are relative. According to Einstein any body in movement is shortened in the direction of the line of motion while the transverse dimension remains the same. If, then, a man is being carried headlong through space with a velocity approaching the speed of light he would be shortened like the man on the left. If he were moving sideways he would be like the man on the right.

The man's image in a plane mirror seems to him symmetrical but reversed. His right hand has somehow got over on his left side and vice versa. Such a transformation as the mirror seems to effect cannot be actually accomplished in ordinary space, but would conceivably be possible in a space of four dimensions.

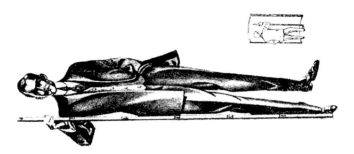

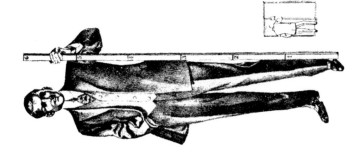

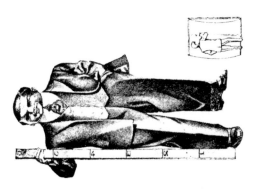

[face p. 20

looks at least six times as high as he is broad. Now you are sure he is cheating—must have some sort of telescoping rod that contracts and expands according to the way he holds it. You point out to him that his measure is unreliable, but to your surprise his gestures seem intended to convince you that you instead are using the elastic rule. You shake your fist in his face—to which he responds with equal indignation—and then you turn to the squatty chap in the other mirror, hoping he will be amenable to reason. But he also measures himself as 72 inches high and 18 inches wide by his own rule. If you try the still queerer-looking fellow in the concavo-convex mirror who is distorted in all sorts of ways you will find that his rule lengthens and shortens and bends just enough to make him as symmetrical a man as yourself. And how can he be otherwise since he is the image of yourself?

You are therefore driven to doubt the invariableness of your own yardsticks. Suppose when you wake up to-morrow everything, including all means of measuring, is twice as big as it is to-day. Could you tell the difference? Would it make any difference? Would there be any difference? Is there any such thing as absolute distance? Are not all measurements relative?

Such questions had from the earliest times occupied the attention of speculative philosophers, but they passed from the realm of metaphysics to the realm of physics in 1886 when Michelson and Morley made their famous experiment on the speed of light in various directions.   Their object was to find out if the ether, the hypothetical medium carrying the light waves, was stationary and drifted back through the earth as the earth moved onward.   They devised an instrument of such delicacy that the stamp of a foot a hundred yards off would be noticeable.   A ray of light was divided into two parts; one half was sent forward and back in the direction toward which that part of the earth where the experiment was made was moving at the time; the other half was sent back and forth across the line of this motion.   But the two rays of light following different routes came back at the same instant and matched up exactly.   In order to correct for any inequality in the instrument, Michelson and Morley turned it around so that the arm that formerly pointed across the line of motion now pointed in the direction of that motion and the other arm pointed across, but that made no difference.   The light travelled with the same velocity regardless of the motion of the earth.

This negative result was just as astonishing as if

you should stand at a certain spot on the bank of a
river half a mile wide and should send out two boats,
one to go up the river half a mile against the current
and then back with the current and the other boat to
go across the river and back.   If both boats should
return at the same moment you would be puzzled to
account for it.   One way of accounting for it would
be that your measurement of the half-mile course up-
stream had been a little short. This was the explanation of the Michelson-Morley experiment given by the
Dutch physicist, Lorentz. He suggested that the arm
of the instrument shortened a trifle as it was turned
from across the line of the earth's motion to the direction of that motion. The amount of shrinkage necessary to compensate for the ether drift would be
exceedingly small.   Besides how could you measure
the change in the length of the arm if the rule you
laid alongside of it altered in the same proportion?
Lorentz's explanation could not be disproved, yet
it was so upsetting to our ordinary ideas of the
stability of matter that it was hard to accept.

Einstein took Lorentz's idea and made it one of the
fundamental principles of his new theory of the universe and then deduced from this theory sundry very
startling conclusions, some of which could be—and
have been—confirmed by experiment. According to

Einstein the size and shape of any body depends upon the rate and direction of its movement. For ordinary speeds the alteration is very slight, but it becomes considerable at rates approaching the speed of light, 186,000 miles a second.   If, for instance, you could shoot an arrow from a bow with a velocity of 160,000 miles a second, it would shrink to about half its length, as measured by a man remaining still on earth. A man travelling along with the arrow could discover no change. No force could bring the arrow or even the smallest particle of matter to a motion greater than the speed of light, and the nearer it comes to this limit the greater the force required to move it faster. This means that the mass of a body, instead of being absolute and unalterable as we have supposed, increases with the speed of its movement. Newton's laws of dynamics are therefore valid only for matter in motion at such moderate speeds as we have to deal with in our experiments on earth and in our observations of the heavenly bodies. When we come to consider velocities approximating that of light the ordinary laws of physics are subject to an increasing correction.

If a person calculates that he is attaining a speed faster than light he will seem to another observer to be moving the other way. That is, any motion above the speed of light is negative motion.   Just as a tourist travelling more than 12,000 miles away from home in any direction will really be getting nearer home the farther he goes.

Such speculations would not have bothered anybody twenty years ago, for then the physicist did not have to handle any cases of such high speeds.   But when radium was discovered it was found that this metal was continuously throwing off particles of negative electricity with approximately the speed of light.   Now if these electrons are not matter they are at any rate the material of which matter is made. They can be detected and counted and tracked and deflected and speeded and weighed.   They are very real things, perhaps the ultimate reality of all things, yet their extreme velocity carries them out of Newton's world and into Einstein's.

INTRODUCING THE FOURTH DIMENSION

Now Einstein's world, as I said before, differs from the world in which we are accustomed to live in many particulars.   It has four dimensions instead of three. One of these dimensions may be time.   Time, too, must be relative, not absolute.   This is even harder to imagine than the relativity of space.

As some schoolboy said : " If there were no matter in the universe the law of gravitation would fall to

WHAT IS MEANT BY DIMENSIONS

No dimensions :
 A mathematical point.
 Has position but no size.
 Represented by a dot.
  Like this

One dimension :
 Has length but no breadtn.
 Made by moving a point along straight in any direction.
 Represented by a line.   Like this—

A plane surface like this page.
 Has length and breadth but no thickness.
 Made by moving a line in a direction perpendicular to
  its length (that is, into the second dimension.)
 Represented by two straight lines of indefinite length
  perpendicular to each other.
 The lines are called axes and are labelled x and y.
 The point where they meet, the origin, is marked O.
                                              Like this—

Three dimensions :
 A solid like a cube.
 Has length, breadth and thickness.
 Made by moving a plane in a direction perpendicular
  to the other two (that is, into the third dimension.)
 Cannot be pictured on paper, but is indicated by three
  axes, x, y, and z, of which x and y are on the plane
  of the page and z is supposed to be stuck up at
  right-angles to the other two   Stick a pin into
  the paper at the point O and you will have the
  third or z axis.   Like this—

Four dimensions :
 Has length, breadth, thickness and extension into a
  fourth dimension, say time.
 Made by moving a cube in a direction perpendicular
  to the other three (that is, into the fourth
  dimension).
 Cannot be pictured on paper, but may be indicated by
  four axes, x, y, z, and t, each at right-angles to the
  other three.   Like this—

More dimensions :
 Any desired number of dimensions can be worked out
  mathematically but with increasing difficulty
  because of the impracticability of diagrammatical
  representation. We can generalize the idea by
  speaking of a " geometry of n dimensions " where
  n may stand for any number whatever from zero
  to infinity.

A line of a given length contains an infinite number of points.
A square of a given size contains an infinite number of lines.
A cube of a given size contains an infinite number of plane
 squares.
A tesseract (four-dimensional cuboid) of a given size contains
 an infinite number of solid cubes.

the ground." Quite so. And what would there be left of space if you took everything out of it, and what would become of time if nothing ever happened? In other words are not space and time merely forms of thought, the framework of ideas, and if so cannot we fix them over to suit our need of new conceptions? As a matter of fact we do. We have constructed by the aid of Euclid and his successors a geometry of three dimensions that works perfectly for all ordinary requirements and if we need a fourth dimension to accommodate these new astronomical and physical phenomena we will build on the necessary addition to our conception of space. There was no use having a fourth dimension so long as we had nothing to put in it. For ordinary earth measurements (geo-metry) such as laying out a town lot we only use two dimensions, length and breadth. We speak of "flat ground" and "water-level" regardless of the fact that all our "straight" lines on the earth's surface are really curves that come back to us after going 25,000 miles or less. It is only when measuring mile lengths that we have to correct for the curvature of the earth in the third dimension. So if, as seems probable, we shall have to make allowance in astronomical measurements for the curvature of the universe in a fourth dimension it will merely mean a little labour to the astronomers and it will relieve their minds of some of their perplexities. There is nothing more mystical or mysterious or " psychical " about a fourth dimension than about the other three. A dimension is simply a measurable direction and we can use five dimensions or n dimensions if we need to.

It does not matter that we cannot " see " a figure in four dimensions even with our mind's eye. Actually we cannot see any figure of more or less than two dimensions : we have to take the others on faith. Nobody can see the mathematician's point because it has no dimensions, no size at all. The schoolboy says, " Let that be the point A," and we let it be although what he is pointing at with his stick is not a point but a vast irregular splotch of white chalk on the blackboard.   So too, we cannot see a mathematical line because it has only one dimension, length and no breadth. But set four lines at right angles to one another and we get a square. This we can really see if the enclosed surface is of a different colour such as a shadow or black print.   Set six squares together at right angles and we get a cube.   This we cannot see in its entirety at one time. All that we see when we look squarely at a cube is a square.   If we look at it from an angle we see what looks like a square

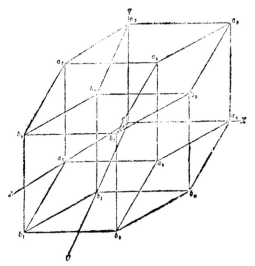

HOW TO DRAW A FOUR-DIMENSIONAL FIGURE

The best way to get an idea of the construction of a cubical solid in four dimensions is to draw a diagram yourself and trace out in turn each of the eight cubes that inclose it. I am indebted to K. W. Lamson of Barnard College for the following sketch and directions:

Draw the four coördinate axes OX, OY, OZ, OU.

Draw the cube $a1a2a3a1a5a6a7a8$ on the three axes XYZ.

Draw parallel to this the cube $b1b2b3b4b5b6b7b8$ on the U axis.

Draw the cube $a1a2a3a1b1b2b3b4$ on the three axes XYU. This is partly drawn already.

Draw parallel to this the cube $a7a8a5a6b7b8b5b6$ on the three axes UYZ.

This completes the figure.

There are four other cubes in the figure besides those indicated:

The cube on the XZU axes $a1a4b1b1a7a8b7b8$ and its opposite $a2a3b2b3a5a6b5b6$.

The cube on the YZU axes $a5a2a7a1b5b2b7b1$ and its opposite $a6a3a8a4b6b3b8b4$.

The figure has:, 16 corners, 32 edges 24 bounding squares 8 bounding cubes.

The heavy line $a1b6$ might be called the principal diagonal and makes an angle of 60 degrees with each of the four axes. It is foreshortened in the sketch, but its real length is twice that of one edge of the cube. Every line except this is on the outside of the four dimensional figure.

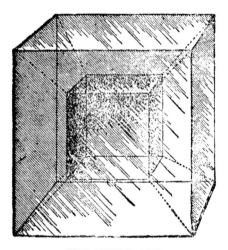

THE TESSERACT

A four-dimensional cube-like solid if transparent and looked at with one eye would appear something like this. But it is obviously impossible to depict a four-dimensional figure on a two-dimensional surface like this page.—From "The Fourth Dimension Simply Explained," Munn and Company, N. Y.

with a couple of lozenges on the sides. The retina of the eye is practically a plane surface, so all we can get is a two-dimensional projection of a solid. Since our two eyes present us slightly different pictures of an object we infer from these its size, shape and distance, but this is guesswork.

Still we have a pretty clear idea of a cube although we have never seen it in its solidity. But the attempt to visualise the hypercube, the four-dimensional figure corresponding to the cube, strains our imagination to the breaking point. Some mathematicians endowed with constructive imaginations of high power claim to have got by long hard thinking some sort of a shadowy and fleeting perception of it, but their visions—if they are not imaginary—do not help out us ordinary folks. But if we cannot imagine— that is, image,—the hypercube we know all about it, even its name. It is called the " tesseract," and it is bounded by eight cubes just as the cube is bounded by six squares and the square by four lines. The tesseract has 24 square faces, 32 edges and 16 right-angular corners.

TIME AS THE FOURTH DIMENSION

Although we find it hard to conceive of a fourth dimension in space we have no such difficulty in case the fourth dimension is time.   In fact, we use this idea all the while and could not get along without it. To fix the position of any event requires four dimensions. For instance, a man is shot. Where? At the corner of 7th Avenue and 42nd Street, New York. This fixes the place by two coördinates crossing at right angles in a plane.   But was it above or below this, on the twentieth floor of the Times Building or in the Subway? Knowing this fixes the third dimension, but we have still to fix its position in a fourth dimension, time.   Was it to-day or last week and what hour?   If then we find out all four we can distinguish this shooting from any that may have occurred in other places at the same time or at other times in the same place.

Or consider this simple illustration : Cut a strip of motion picture film into its separate scenes and pile them up in order till it is as high as it is broad. You have then a cubical event.   Two dimensions of the cube are spatial; the third dimension is essentially temporal, although in a spatial form. If one of the films from the middle of the pack represents the present then the films below represent the past and those above the future.   The people on the picture you picked out know only of the scene there depicted though they may have a fading memory of the past and a dim anticipation of the future. But to you who are outside of the film pack all the scenes are equally visible. They are all present to you. This is the way most Christians have conceived of God, as one to whom past and future form one eternal present, so He sees simultaneously all things that have been, are or will be.

If our pile of film were made up of snapshots taken one a day throughout a man's life we should see at one glance his growth from babyhood to boyhood, to maturity and old age. We could turn the leaves of his life backward or forward as we will. Some day perhaps we shall have stereo-movies, scenes in three dimensions with time as the fourth.

This idea of time as a fourth dimension is not a new one. In 1754 d'Alembert defining " dimension" in the Encyclopedia, wrote : " A brilliant man of my acquaintance believes that one may regard duration as a fourth dimension." In 1903 Minkowski worked out the idea in mathematical form. H. G. Wells, always quick to catch up a new scientific theory to use as a plot for a story, wrote in 1895 of "The Time Machine," a vehicle by which a man could travel back and forth in time as he can travel east and west in a motor car. In this he visits the future and finds mankind split into two species, a subterranean

C

working class living on—literally—a pleasure-loving
leisure class.

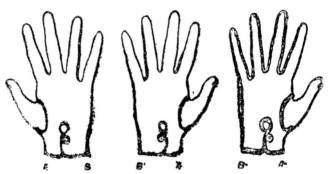

In space of three dimensions we cannot make a right hand
glove and a left hand glove look the same no matter how we
turn them around. But if we turn one glove inside out it will
match the other except that the lining now appears on the
outside.

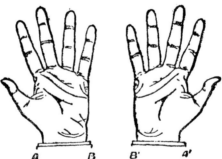

Our two hands cannot be turned inside out so as to look the
same in three dimensions, though they might in four
dimensions.

In "The Plattner Case" Wells tells of a chemical
professor who was by an explosion knocked into—not

the middle of next week as we commonly say—but
into the fourth dimension of space.   Ten days later
he was knocked back again into our world but the
only evidence of the truth of his story was that his
heart beat on the right side and he was left handed
and otherwise reversed in a way that would be impossible in a space of three dimensions. We can turn
a glove inside out in three dimensions and so make it
just like its mate of the other hand, but we cannot
turn a solid inside out except in four-dimensional
space.

In another of his " Thirty Strange Stories " Wells
tells " The Story of Davidson's Eyes." While Davidson was working in his London laboratory a light-
ning shock so affected his eyesight that he could not
see the familiar objects about him which he could
feel but looked instead at a South Sea island on the
opposite side of the globe. This might be possible in
a curved space of four dimensions although Wells
professes to pooh-pooh such an absurd suggestion
while he ingeniously insinuates it. George Macdonald
in his fantastic romance " Lilith " also introduces
the fourth dimension.

Points that are far apart if measured in three dimensions may be close together in the fourth. We
can readily understand this if time is the fourth di-

mension for events can happen at the same instant
though thousands of miles apart.   But it is not
impossible to conceive of the fourth dimension as
spatial instead of temporal if we approach the problem from a simpler standpoint.   Let us think of
ourselves as living in a "Flatland" of two dimensions with no thought of a third.   There yet survive
in enlightened America individuals who believe that
"the sun do move" and who deny that the earth is
"round like a ball."   That is, they do not recognise

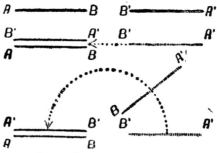

By movement in one dimension we cannot make the lines AB
and BA coincide for if we drag BA straight on to AB the ends
will not match.   But if we swing AB around through the second
dimension we bring it on AB so the letters correspond.

the curvature of the earth in the third dimension.
But if such an individual were to travel in a
"straight" line westward over the "level" land
and water he would, much to his surprise, come
back to his starting point which he had left 25,000
miles behind him.

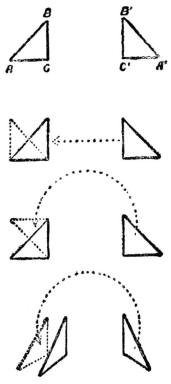

In space of two dimensions, such as a table top, we cannot bring these two triangles into the same position. If we drag one over on to the other (movement in one dimension) they will not fit together. If we swing one triangle around (movement in two dimensions) they still do not fit. But if we take one triangle off the table and turn it over (movement in the third dimension) we can then lay it by the side of the other and they will match perfectly.

A WORM'S-EYE VIEW OF THE WORLD

Suppose yourself a worm—the Bible says you are anyway—and crawling around on a sheet of paper. With your vermicular mind you doubtless would take a superficial view of the universe and find it as impossible to imagine a third dimension as man does

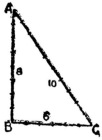

a fourth. If in the course of your crawling you came across a triangle you might—if you were a measuring worm—pace it off and find that the distance from A to B was 8 inches, from B to C was 6 inches and from this data, if you knew the law of the hypothenuse, you might calculate that the distance from A to C was 10 inches. On measuring it you would find your prediction verified, and so gain perfect confidence in your plane geometry.   But unbeknown to you, poor worm with your eyes fixed on the paper, some man may have picked up the sheet and crumpled it up or rolled it over so that A and C are then only one inch apart in the third dimension. The worm is right when he thinks the distance between these points is 10 inches; so is the man right when he says it is one inch. It depends on the point of view.

Now in Einstein's view something of this sort happens to our three-dimensioned space when matter gets into it. We know for instance that if you divide the circumference of any circle by the diameter the ratio figures out as 3.1415+. It has been calculated to 707 decimal places but we can dispense with the rest of them and call the whole thing Pi for short. Write it in Greek as π and it looks more learned. Now if you place a heavy particle, say a lead bullet, in the centre of a circle the ratio of the diameter to the circumference according to Einstein, becomes a little less than Pi, for the circle has been warped, so to speak, into the fourth dimension by the strain of gravitation. The difference in such a case is too small to be measurable by any known means, but it is supposed to be an actual, not an imaginary, deviation from the geometrical law.

Now the sun being a big heavy body must extend its gravitational strain for a considerable distance around and a ray of light passing through this crumpled up space would not be able to pursue a straight course. And, according to the eclipse observations, it does not.   Light like everything else follows " the easiest way " and this is not always the straight and narrow path. A river takes the easiest, not the shortest, way to the sea and this leads it through many meanderings.   Most of us, I suppose, have a mental image of Newton's gravitation as a sort of rope by which the sun pulls the earth into its orbit when it is disposed to fly off on a tangent. But from Einstein's viewpoint we should rather think of the earth as picking its way as best it can through a space-and-time combination that has been strained and distorted by the power of the sun. I visualise Einstein's solar system as a spider web with the sun in the middle like the spider and the planet like flies trying to get around through the tangled strands. But it is more complicated than that for each planet has its own lesser web of radiating influence to drag about with it wherever it goes.

Newton's idea is simpler, but unfortunately light at least seems to follow Einstein's law, not Newton's. That is why Einstein is such a troublesome fellow. If he would confine himself to metaphysical speculation nobody need bother about those strange notions of his.   But when he points how they can be proved and then British astronomers and American physicists find things according to his deductions he cannot be ignored. The man does not seem to have that decent respect for the opinions of mankind that leads most of us to limit our logic to the sphere of common sense. When he gets an idea in his head he follows it wherever it leads him even though he bumps up against Euclid and Newton and the rest of us. For instance if you admit the second of his two fundamental postulates, that the speed of light is constant, regardless of the velocity of its source, you are irresistibly led— unless you let go of his hand somewhere on the way— to the conclusion that time is a local affair; that there is no way of telling by light signals whether two clocks at a distance are keeping the same time, or whether two events at different places occur simultaneously. You could not tell this even if you could shoot a watch from one place to the other with the speed of light, for no matter how many seconds, or years, the watch might be on its way it would register the same time. If instead of a watch a man could travel at that speed he would not grow old on the way. According to Einstein no man, watch, or any other material thing can travel with the speed of light, for it would require an infinite force to give the smallest particle such a velocity. But let us suppose that a hollow projectile holding a man, such as Jules Verne and Wells used on their voyages to

the moon, should be sent off into space with a velocity
one twenty-thousandth less than light. If at the end
of a year the projectile should be caught like a comet
by the gravitation of some star and be swung around
and sent back to earth, the man on stepping out of
his shell would be two years older but he would find
the world two hundred years older. This would be,
as Professor Langevin suggests in *Scientia*, 1911, an
interesting way to study history, but it would be
risky, not to say impossible. Still French scientists,
like Napoleon, have no place in their dictionaries for
so stupid a word as "impossible" and M. Esnault-
Pelterie has figured out that a thousand pounds of
radium would be sufficient to carry a man to Venus
in 35 hours if a hollow projectile could be fitted up
like a rocket with the radium in the rear sending out
a rapid fire of electrons.

TURNING TIME BACKWARD

To loosen up our conventional ideas of the fixity of
time and space we may accept the aid of the scientific romancers. Camille Flammarion, the famous
French astronomer, wrote a fantastic little book
called "Lumen" which tells of a man who died in
1864. His soul flew straight to its heaven which was

one of the planets of Alpha, the largest star in the constellation Capella. Here he found the benevolent inhabitants of that sphere, who were endowed with superhuman powers of sight, watching with great distress the bloody scenes of the French revolution of 1793, and wondering how it would come out. To the visitor from the earth this was an old story, to the people of Alpha Capella, it was a present spectacle, for the distance of the star was such that it took light 72 years to travel from the earth, so they were 72 years belated in their observation of current events on our planet.

The spirit of the defunct Parisian, having the power of flying through empty space at any speed he chose, found that he had thereby also acquired control of time and could hasten, retard, stop or reverse the course of events at will by simply varying his speed. If he remained stationary, scenes on the earth would unfold at their normal rate and in regular order. If he travelled away from the earth with the speed of light everything seemed to stand still. If he travelled faster than light he overtook the rays that had left the earth farther and farther back in the past so he saw through them events in the reverse order. For instance when he looked down on Waterloo he saw the battle-field strewn with corpses and Napoleon

walking toward Waterloo backward pushing his
horse by the bridle. This is how the battle looked to
the interspatial observer.

When my sight was sufficiently habituated to the
scene, I perceived some soldiers coming to life out of
the eternal night, and by a single effort standing up.
The dead horses revived like the dead cavaliers, and
the latter remounted them. As soon as two or three
thousand men had returned to life, I saw them form
unconsciously in line of battle. The two armies took
their places fronting one another, and began to fight
desperately with a fury that one might have taken
for despair. As the combat deepened on both sides,
the soldiers came to life more rapidly.

At each gap made by the cannon in the serried
ranks a group of resuscitated dead filled up the gaps
immediately. When the belligerents had spent the
whole day in tearing one another to pieces with
grape-shot, with cannons and bullets, with bayonets,
sabres and swords—when the great battle was over,
there was not a single person killed, no one was even
wounded; even uniforms that before it were torn and
in disorder, were in good condition, the men were
safe and sound, and the ranks in correct form. The
two armies slowly withdrew from one another; as if
the heat of the battle and all its fury had no other

object than the restoration to life, amid the smoke of
the combat, of the two hundred thousand corpses
which had lain on the field a few hours before.
What an exemplary and desirable battle it was !

Another literary curiosity on the same theme is
" Ignis" by Comte Didier de Chousy.   This tells of
certain engineers who attempted to utilize the internal heat of the earth by running the waters of a lake
into a deep boring.   The result was an explosion that
blew off a piece of the planet.   But the passengers on
this artificial asteroid on looking down through their
well at the earth they had left could see the lake and
city undisturbed and watch themselves at work as
they were before the place blew up.   The explanation
was that this fragment of the earth was projected into space more rapidly than the speed of light and so
was catching up with the rays that had gone out
before the explosion; these rays, of course, carried
the picture of earlier scenes.   But Einstein would
say that this story—as we might ourselves have suspected—must be fiction for according to his theory
the speed of light is the absolute limit of motion, the
infinity of velocity, which no material body may ex-
cel or attain.   He does not however say anything
about the possible speed of a disembodied spirit such

as Flammarion employed in his imaginary explora-
tion of space.

THE METAPHYSICS OF THE MOVIES

But from such fantasies we can see that the order
in which we view events depends upon how fast and
in what direction we are moving, and the past and
future may be reversed to our vision.   This is easily
made apparent by means of motion pictures. If the
film is reeled off in the wrong direction the action is
reversed.   So we see divers rising gracefully out of
the water and landing on the spring board.   Newly
hatched chickens dismayed at the sight of this un-
friendly world calmly took themselves back into
their broken shells which close in upon them. When
we have come to the close of a perfect Thanksgiving
Day the obliging operator may give us an encore of
the dinner reversed by running his machine backward.
Then we see pieces of turkey politely picked out of
the mouths of the diners with their forks and re-
placed upon the plates. When these are passed back
to the carver he puts the slices neatly in their places
and the fowl is then sent back to the oven to be un-
roasted. The cook then sticks on the feathers. The
hired man carries the turkey out to the chopping

block where with one swift stroke he restores the head and the fowl runs off backwards. This is just as correct as the ordinary order. The sequence of events is the same. Cause and effect are linked together as firmly as before, only they have exchanged places. A scientist knowing nothing of our world except from waching such reversed motion pictures might deduce from them the same consistent and logical system of natural laws that we now have although some of them, for instance, the second law of thermodynamics, would be reversed in form.

The motion-picture man has also the power to alter the speed of the passage of time as he will by turning the crank faster or slower. Sometimes he is quite too careless in the way he employs this prerogative. If he is behind time on his schedule he will rush through a lazy siesta scene in a Mexican plaza with all the fury of a Mack Sennett farce. But this telescoping of time can be used to advantage as when he shows us the growth of a plant, the unfolding of its flower and the ripening of its fruit, all in fifteen minutes. On the other hand motion may be slowed up by taking twice as many pictures a minute as usual and projecting them at the ordinary rate. For instance, if it is a dog jumping up to grab a piece of meat from his master's hand, we see the dog rise slowly from the ground and, while poised in mid-air, eye the meat carefully to select the best point of attack, then deliberately take it between his jaws and gradually descend. Now notice that this is just as true a picture of the dog's jump as any other. The movie man has simply expanded time measurements as he expands space measurements when he shows us a close-up. A close-up with a face covering a sixteen-foot screen is just as true as a smaller picture. It is what we should always see if the lens of our eyes were a bit more convex. We look through the small end of an opera-glass and objects seem magnified. We look through the large end and objects seem minified. This is not an illusion. The opera-glass *does* actually enlarge or reduce *what we see.*

So too time intervals can be lengthened or shortened. Take a dose of hashish—no, don't—I should say, if you did take a dose you would find that your perception of duration was prolonged. If while under the influence of the drug you drop a book it will seem an hour getting to the ground. De Quincey describes such experiences in his "Confessions of an Opium Eater." But without entering into such abnormal states we all know by everyday experience how time flies or lags according to the number of our sensations. Bergson's philosophy is built upon the distinction between the idea of duration as experienced by all of us, and the idea of time as established by the physicists for comparative measurements.

> We live in deeds not years; in thoughts not breaths;
> In feelings not in figures on a dial.
>
> *—Festus.*

For all we know an ephemeral insect that dies in a day may live a longer life than a Galapagos turtle that exists for two centuries.

What Mark Twain said about classical music applies also to science: " It is not so bad as it sounds." The thing that the chemist calls " sodium chloride " other folks call " salt"—and so does he when he is off duty. Don't let the scientist bluff you by his polysyllabic propensity. Just try to see what he means by such language. Now what these new-fashioned non-Euclidean geometricians call "the four-dimensional space-time continuum" is essentially the same system of reference as you have used ever since you could toddle. Minkowski did not invent it. Everybody thinks that way unless he is an idiot. Each one of us has had to build up his own phil-osophy of the universe long before we went to school, mostly before we could talk. We had to study geometry while we were in our cradles—worse than that we had to work out a practical system of geo-

D

metry for ourselves without the help of Euclid or anyone else.   We had to excogitate a system of relationship between the sights and sounds and touches that came to us before we could get along in the world.   Probably we all solve this riddle of the universe in about the same way although since there is no way of directly comparing notes we cannot be sure about that.

THE EGOCENTRIC THEORY OF THE UNIVERSE

But the framework that we construct to hold everything outside of ourselves is essentially of the following form :

You are the centre of your universe.   Everything and every event that you are considering is related to you here and now.   Starting from this, your point of place and time, you imagine eight straight lines stretching out toward infinity in eight directions as divergent as possible.   These lines—call them destinations or directions or dimensions or coördinates as you please—consist of four opposing pairs, right and left, up and down, forward and back, future and past.   Somewhere along or between these four dimensional lines that cross in your brain you can find a place for anything that you need ; your pencil, the discovery of America, the sun, and next Friday.

You can connect up all these things by lines which may represent changes, that is the tracks of movements in space and time. To connect the pencil in your hand with the discovery of America you would have to count back 428 years on the time line and measure off on the east-west and north-south lines whatever distance you may be from San Salvador—not to consider the motion of the earth.

Anything that exists, that is to say, persists, is moving along the time dimension at what appears to be a uniform rate. Of course you can, if you like, conceive of time itself as a stream flowing through things. Since all motion is relative, that way of looking at it is just as " true " as the other. But it is simpler and more sensible to think of things moving through a stationary time just as we think of them moving through a stationary space. A material point that is at rest, such as the dot of an i on this page, (we continue to disregard the motion of the earth) is not moving about in space but is moving forward in time. Its track then is a straight line along the time dimension. That is, a material point is a line in the fourth dimension. If you move the page to the right the forward movement of the dot of the i in the time dimension is combined with the sideways motion in a single slanting line. If you move the page simultaneously upward, rightward and backward the track of the point is a line combining the movement in all four dimensions. Such a track of a point moving through space and time is called its "world-line." It is a continuity of one dimension. Any event is the point of intersection of one or more such world lines and we can never observe anything except such intersections. That is to say, everything happens somewhere and sometime.

A picture flashed on a cinema screen has three dimensions. It is, say, 10 feet long and 6 feet high and lasts 1-16 of a second, but it has no thickness. A man necessarily has four dimensions. He may measure from 24 to 72 inches in one dimension, from 8 to 18 inches in the second, from 4 to 9 inches in the third and 70 years in the fourth.

After all, the idea of the relativity of time ought to be easier to accept than that of space for it is in accord with experience instead of contrary to it. We drop off to sleep and wake the next instant if we credit our personal perceptions. Why should we believe the sun and the clock in preference to ourselves?

Bergson bases his whole philosophy upon the distinction between *duration* as it is felt by the individual while he is living through it and *time* as it is employed by the physicist in his calculations. The latter conception, physical time, is, as Bergson says, a mere invention of man and virtually a fourth dimension of space, so he concludes :

> To sum up; every demand for explanation in regard to freedom comes back, without our suspecting it, to the following question : " Can time be adequately represented by space?"   To which we answer : Yes, if you are dealing with time flown; No, if you are speaking of time flowing.*

Past and future are alike to the physicist, differing only in direction, like east and west.   But to the living person they are altogether different things. For man rolls up his past, as a tourist his rug, and carries it with him wherever he goes.   That is why Well's " Time Machine " and the reversed reels of the movies are so funny.   There is nothing absurd about running a wheel backward but there is about running a man backward.†   The physicist feels no reluctance about turning the stream of time backward for all physical phenomena are reversible under the proper conditions. If we interpret the universe as merely matter in motion and imagine at a certain instant that every individual particle reverses its motion and

*Bergson : " Time and Free Will," p. 221.

†Bergson in his " Laughter " traces all humour back to this fundamental absurdity of making a man act mechanically.

goes in just the opposite direction at the same speed, then the whole history of the world would be reënacted in the opposite order and the earth would return to its primeval nebulæ.

In Wells's story, "The New Accelerator," a professor invents an elixir that speeds up the rate of living a thousandfold. A person taking a dose of it sees people as wax figures apparently motionless in the midst of violent action. Falling objects seem to stand still in he air. The music of a band is reduced to "a low-pitched wheezy rattle" or "the slow muffled ticking of some monstrous clock." But in compensation for this the accelerated drug-fiend could watch at leisure the slow flapping of a bee's wings.

But even Wells with his seven-league-boots imagination finds it difficult to keep ahead of the march of science. What he then saw only with his mind's eye we can actually observe. By moving the accelerating lever on your phonograph toward the S end of the scale you can slow up the tune and lower its pitch until it becomes inaudible as music. The new Pathe ultra-rapid camera can take pictures at the rate of 160 to the second. When these are projected on the screen at the usual rate of 16 to the second all movement takes place ten times slower than in actual life. This gives opportunity for the study in detail of the action of a ball-player pitching a curve or of the wing motion of a humming bird or of the splash of a marble falling into water or of the flight of a bullet. We can magnify motion or minify it as much as we will. The cinematograph owes its origin to the desire of Senator Leland Stanford to study the movement of a horse's legs so as to find out why one racer went faster than another.

Such playful flights of the scientific imagination as Wells and Flammarion indulge in and such freaks of projection as the camera man amuses us with are of use to those of us who find difficulty in translating a mathematical formula into terms of everyday life. There is no better place to study metaphysics than in the world of the flickering screen, for there man has complete control of time and space. He can enlarge and reduce any object. He can hasten, retard or reverse any action. He can throw upon the screen at the same time events happening months and miles apart. Therefore to those of us who have had the advantage of an education in the movies, Einstein's ideas of the relativity of time and space do not seem startling or inconceivable.

Kant not only conceived the possibilty of more than three dimensions but believed in the probability of it. His argument is based on greater insight into the intentions of the Almighty than we of this day would claim :

"If it is *possible* that there be developments of other dimensions in space, it is also very *probable* that God has somewhere produced them. For His works have all the grandeur and variety that can possibly be conceived."

In this temporal, spatial and material world of ours reality requires that the four dimensions should hang together.   But at an infinite distance from all matter this fourfold combination would be dissolved into a three-dimensional space and a one-dimensional time.   In that extra-mundane realm time ceases to flow, gravitation no longer drags downward, matter is non-existent, light is immovable and change is impossible.*   Thus the new mathematics leads to a state curiously like the conventional conception of heaven.

We talk as our forefathers did about " the ends of the earth " but we know that one might start from his home and walk forever in any direction without coming to an end of it.   But though the earth's sur-

---

*" We can thus say that all those paradoxical phenomena (or rather negations of phenomena) which have been enumerated above can only happen after the end or before the beginning of eternity." (De Sitter).

face is infinite in the sense of endless, yet one never can get more than 8,000 miles away from home where'er he may roam. If a man stood on the top of the highest mountain on earth and aimed a level gun in any direction, the bullet, if it could be given sufficient velocity to overcome the influence of gravity, would go around the world and hit him in the back of the head. Or if light were sufficiently deflected by gravitation to follow a level line around the earth—another absurd assumption—the man looking through a level telescope in any direction could see how his hair was combed in the back. Such happenings, though impossible, are not inconceivable but are logical consequences of our knowledge that the world is round and that what we call straight or level lines as measured on plain or sea are really great circles around a centre four thousand miles below.

Now is it not also conceivable that the lines we call straight in astronomical space may also have an imperceptible curvature in some unknown fourth dimension? If this curve is closed like the circumferences of the earth a ray of light pursuing a straight course in a certain direction might eventually return upon its track, even though not reflacted or reflected by the matter it passes through or by. A telescope of unlimited power pointed into space at a tangent might then show the observer his own back, if light were transmitted instantaneously, but, since it is not, and since the curvature of space, if there be any, is exceedingly minute, what the observer would see, assuming that the earth had come back to its former position, might be the scenes of some geological age millions of years ago.

NON-EUCLIDEAN GEOMETRY

The idea that space may itself be curved and that the axioms and assumptions on which our geometry since the time of Euclid have been based, may not be absolutely and exactly and eternally and universally true has been diligently studied during the last fifty years. The Russian Lobatchewsky, the Hungarian Bolyai and the German Riemann have developed systems of geometry by starting from premisses the opposite of those of Euclid and these systems are just as logical and consistent with themselves as the ordinary or Euclidean geometry. These non-Euclidean geometries were at first commonly regarded as mere freaks of the mathematical imagination, but they have already proved valuable in leading to a reconsideration of the fundamental principles of our thinking and, if Einstein is right, they may be necessary to explain physical phenomena.   It is hard for the mathematician to discover anything useless.   A distinguished American mathematician in announcing a new theorem exclaimed : "And thank Heaven, no possible use can ever be found for it." But, whatever it was, he made a rash boast for nowadays the mechanic treads on the heels of the mathematician and uses imaginary quantities, actual only in the fourth dimension, like $\sqrt{-1}$, in figuring out the winding of his dynamo.

Readers whose mathematical faculty is weak or undeveloped and who like something concrete with " human interest " in it will find what they want in " Flatland by A. Square," a book published in London in 1891. The author, who turned out to be the Reverend Edwin Abbott, tells of a land in only two dimensions. The ruling class consisted of polygons, the bourgeoisie of squares and equilateral triangles, the lower class of isosceles triangles of narrow base, while the criminals had more irregular forms and the women were mere needles. Since all were confined to a surface, four lines set in a square made a tight prison. The inhabitants of Flatland, even the aristocratic and intellectual individuals who had so many sides as to be almost circular, could not conceive of a third dimension from which a person like ourselves could look down and see at a glance the insides of their houses, their safes and their bodies, just as a being in the fourth dimension could see the insides of ours. The narrator, that is, A Square of Flatland, visits as a missionary the land of two dimensions where all the people lie in a line and refuse to believe in anything outside it, and finally he encounters and endeavours to convert a solitary point of no dimensions but finds him, as we should expect, an incorrigible solipsist.

We should all of us have been familiar with the fourth dimension for years if Slade had not turned out a trickster. Slade was an American medium—the original of Browning's " Mr. Sludge "—who fooled Professor Zöllner by giving him what purported to be experimental evidence of the fourth dimension. Zöllner was a distinguished German physicist, Professor of Astronomy in the University of Leipzig, old, near-sighted, pre-disposed to spiritualism, and unskilled in legerdemain. Any proofs that Zöllner asked for, Slade was usually able at the next

In space of one dimension (a straight line) there could be neither bend, loop nor knot in a string.

In space of two dimensions (a flat surface) a double bend could be made in the string but no loop or knot could be made.

But if we raise one string (into the third dimension) and lay it over the other like this:

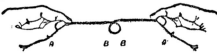

We get a loop but cannot form a knot without using the ends.

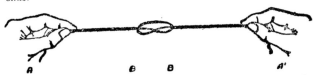

A knot like this cannot be made in a string so long as the ends are held by the hands. But if we could use a fourth dimension we could tie such a knot as easily as we made a bend by the use of the second dimension and a loop by the use of the third. If such a knot could be tied in a string so held it would be experimental evidence of the existence of four-dimensional space.

séance to produce.   All the things one might do in
four dimensions but could not do in three were forth-
coming by the obliging spirits whom Slade had at
call.   Zöllner tied the ends of a string together and
sealed them on the table top, letting the loop hang
down under the table out of sight.   He then asked
to have a single knot tied in the string and the spirits
tied four.   Zöllner also reports that the coins he put
into a sealed box were taken out and writing produced inside sealed slates.

On the basis of these experiments Zöllner wrote a
volume on " Transcendental Physics " to prove the
existence of another world in the fourth dimension.
But when Slade tried his tricks in London he was
caught at them by Professor E. Ray Lankester.   He
was convicted of deception with intent to defraud in
the Bow Street Police Court and sentenced to three
months' imprisonment with hard labour.   Nowadays
the apparatus for Slade's famous slate-writing trick
can be purchased at any conjurer's shop.

It is vain to expect anything scientific to come out
of the séance room where the alleged phenomena are
not reproducible under specified conditions but appear only occasionally and under circumstances prescribed by the medium who always may be and often
is proved to be a sleight-of-hand—or sleight-of-foot—

performer.   The fourth dimension which Einstein and other scientists are now considering is not conceived of as the abode of departed spirits, a spare room for ghostly visitants, but merely as a new factor in a mathematical formula.   It offers us no hope of ever being able to take coin out of a closed safe or put coin into an unopened cocoanut, but it does promise to explain certain optical phenomena which, though rare and minute, are yet open to the observation of anybody, be he sceptical or credulous.

SOME SIMPLE EXAMPLES

Lisbon lies nearly straight east of New York but when a ship captain wants to go to Lisbon he does not sail straight east but sets his course a little northward in the beginning and a little southward toward the end and so gets there quicker than if he had followed a line of latitude.   Draw his course on a flat map and you would think he was taking a roundabout route, but trace it on a globe and you will see that he is following a great circle, the geodetic line, which is the shortest distance between any two points on the earth's surface.

An airman looking down on a rocky, hilly, woody country sees it as a flat plain and if he watched a hunter returning home with his bag of game would wonder that he did not go straight instead of wandering around in such an irregular way. Yet the hunter, being tired, is taking what is for him the shortest way home as he dodges rocks and circumambulates the hills. The easiest way is the shortest way.

A river in its desire to reach the sea always takes the shortest possible way. Its meanderings are not meaningless but determined by a law as rigid as a law of geometry, that is, the law of gravitation which prevents the river from taking a short cut over the hill.

If you look at a landscape over a heated plain or bonfire or through uneven glass you will see that the image is distorted and confused because the rays of light are refracted and entangled as they pass through this unequal medium. Yet each ray is going just as straight as it can toward your eye.

Now to such familiar cases where a ray of light is bent out of its straight course by the uneven density of the air or glass through which it passes Einstein has added another and unsuspected effect, namely, that light is likewise deflected in passing through a strong gravitational field such as the vicinity of a large body like the sun.

It has long been known that the displacement of the earth in space and time (that is to say, its motion) causes an apparent displacement of the stars in space.

The astronomer does not point his telescope straight at a star. If he did, he would not see it, for, owing to the forward motion of the earth, the telescope

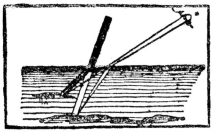

Everyone knows that a ray of light is bent out of its straight course as it passes from the air into a denser medium like water or glass, and that this deflection apparently shifts the position of the object from which the light comes. Einstein's theory and the British eclipse observations prove, what was not known before, that a ray of light as it passes through the gravitational field of a large body like the sun, is also percepibly bent out of its straight course and likewise makes an apparent shift in the position of its source, the star.—From Black & Davis' "Practical Physics." Published by The Macmillan Company.

moves out of range of the rays that otherwise would have reached it.

If you have ever tried to shoot a bird on the wing or, better, a prairie-dog from a train you will get the idea. Or, if you have not had this experience, you have doubtless watched the raindrops running down

E

a car window and have noticed that when the rain is falling straight down the drops strike the pane on a slant when the car is moving forward. The faster the car moves the greater the deviation from the perpendicular. If the train runs backward the rain-streaks slant in the opposite direction. If then you should be asked to point out the direction of the cloud from which the rain is coming you would—unless you knew and made allowance for the movement of the train—point in a line with the streaks on the pane, sometimes backward, sometimes forward, but not straight upward where the raincloud really is.

Now the astronomer is on a moving train, the earth, which is rushing around a ring about 186,000,000 miles across. Consequently every star appears to wobble around in a little ellipse and the astronomer has to aim his telescope, now on one side, then on the other, of the real position of the star in order to bring it on the cross-hairs of his object glass. This apparent displacement of the stars known as " the aberration of light " was explained by Fresnel in 1818—to everybody's satisfaction until recently— on the assumption that all space is filled with an im-movable medium, the ether, which transmits the rays of light in straight lines in the form of wave motion, and that the earth moves through the ether without displacing it, somewhat as an airplane moves through still air. But the aviator knows how fast he is moving by the current of air streaming back in his face. Why then, since the ether is in perfect repose, could we not determine the absolute motion of the earth through space by measuring the drift of the ether as it streams through the pores of the earth? Light appears to afford us a means of measuring such a drift of the ether through matter, if there be such. Since light is conveyed by the ether we should naturally expect it to take less time to travel a certain distance if the receiving instrument is carried toward the source of the light by the earth motion than if it is being carried away from it. This question was put to the crucial test by two American physicists, Michelson and Morley, who devised an instrument so delicate that it could detect differences of one-25,000,000th of an inch in the path of a light ray. But although this delicacy was ten times greater than was necessary to detect the ether drift, if there were any, no evidence of such drift could be discovered.

THE ECLIPSE OBSERVATIONS

In the history of science the year 1919 is likely to be known, not as the year of the overthrow of the

German Empire, but as the year of the overthrow of
Newton's law of gravitation.   The British astronomers who went to Africa to observe the eclipse of the
sun May 29, 1919, came back with the proof that a
ray of light passing close by the sun is bent out of its
straight course.   The photographs taken during the
six minutes when the sun was shadowed show the
surrounding stars in different positions from where
they are seen when the sun's disk is not in their
midst.   This is the second time that Einstein has
scored over Newton.   The first was in regard to the
orbit of Mercury.   If the sun and Mercury were alone
in the universe the planet, according to Newton's
law, would revolve forever around the sun in the
same elliptical track.   But the presence of the other
planets makes Mercury deviate from this regular
route, so the ellipse it describes is never quite the
same but slowly shifts around so that in the course
of centuries its longer diameter would be pointing in
a different direction.   Calculating by Newton's law,
the influence exerted by the other planets, astronomers found that it would shift the orbit of Mercury
532 seconds of arc in a century.   But when they took
observations on Mercury they found that its orbit
was shifting at the rate of 574 seconds.   The discrepancy between observation and theory, 42 seconds, is

thirty times greater than could be accounted for by errors of instruments or observation. But according to Einstein's theory, if the sun and Mercury were alone in space with no other planets interfering, the orbit of Mercury would not remain the same, but would advance at the rate of 43 seconds a century. This, as the reader will observe, is in substantial agreement with the discrepancy which has for two centuries puzzled astronomers, since it was inexplicable on the Newtonian theory.

The electro-magnetic theory of light, thought out by Clerk Maxwell forty-five years ago, has proved to be an excellent guide to research and led to many practical applications, such as wireless telegraphy. According to this theory the miles-long Marconi waves, the infinitesimal waves that we feel as heat or see as light, and the still more minute waves of the X-rays are movements of the same sort, though differing in length, and all travel at the same speed in space of 186,000 miles a second. It was one of the implications of Maxwell's theory, though it was not perceived until later, that light and all such waves must exercise a certain pressure upon a body against which they strike, just as a jet of water from a fireman's hose pushes against the side of a house. The pressure of light is so exceedingly slight that it had never been noticed, but it has been actually detected and measured by Professors E. F. Nicols of Yale and G. F. Hull of Dartmouth. The sunshine falls upon the earth with a force of 160 tons. Both theory and experiment have shown that a beam of light has inertia or mass, that is to say, a beam of light pushes like a water jet, and it has now been proved, by the eclipse expedition, that the pull of gravity deflects a beam of light as it does a water jet. That is to say, a beam of light has weight, is attracted by gravity. This deflection of a beam of light by gravity is extremely small, but photographs taken during the recent total eclipse of the sun show that the star beams that passed near the sun are bent out of a straight path.

A better illustration of the eclipse observation than

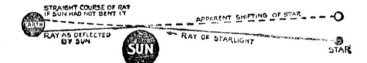

The eclipse expedition found that the stars seen about the sun appear slightly shifted from the positions they occupy on a map of the same region of the sky when the sun is not in their midst. This shows that a ray from a star is refracted or bent as it passes close to the sun and confirms Einstein's theory that light is affected by gravitation. The observed angle of deflection agrees closely with that predicted by Einstein but is twice as great as that required by Newton's theory of gravitation. In this diagram of course the angle of the deflected ray and the size of the sun and earth relative to distance are greatly exaggerated.

I could word is given by Sir Oliver Lodge in his interesting article on " The New Theory of Gravitation" in the *Nineteenth Century* of December, 1915, from which I therefore quote :

Take a fine silk thread of indefinite length, and stretch it straight over the surface of a smooth table or floor. Imagine a star at one end of the thread, and an eye at the other; and let the thread typify one of the rays of light emitted in all directions by the star, viz. the ray emitted in the direction of the observing eye.

Now take a halfpenny (or an American quarter), place it on the table close to the thread, so that the eye end of the thread is ten feet away; and then push the halfpenny gently forward, till it has displaced the thread the barely perceptible amount of one thousandth of an inch. The eye looking along the thread will now see that the ray is no longer absolutely straight; in other words, the star whose apparent position is determined by that ray will appear slightly shifted. The scale is fixed by the size of the halfpenny, whose diameter, one inch, is used to represent the Sun's diameter of 800,000 miles. The ten-foot distance between eye and Sun practically supposes that the eye is on the Earth, which would be a spot one hundredth of an inch in diameter, or about the size of this full stop.

As for the distance of the star, at the other far end of the thread, that does not matter in the least : but, on this scale, it may be interesting to note that one of the nearest stars, about eight light-years away, would require the thread to be a thousand miles long.

The ray is now bent or deflected as it passes the neighbourhood of the Sun on its long journey, so that it is out of place one thousandth of an inch at a distance of ten feet; and the effect of this tilt of the ray, upon the observer, is to make him just able to see a star upon the Sun's ' limb ' when it is really behind it, or to make him see a star slightly further off the ' limb ' or rim of the

Sun than it really is. The shift of one thousandth of an inch at a distance of ten feet corresponds to an angle of one and three-quarter seconds of arc, which is just the optical shift that actually ought to occur, according to Einstein, when a ray from a star nearly grazes the Sun's limb on its way to a telescope; and this is the optical shift which we now know does occur. That may be taken as the definite result of the recent eclipse observations. The effect, both in magnitude and direction, had been predicted four years before, on the strength of a mathematical investigation, by Professor Einstein.

The images of two stars, one on each side of the sun's disk, will apparently be crowded a little apart when the sun comes between them. A star that would be just eclipsed by the edge of the sun's disk if its rays came straight may still be visible since the rays are curved. In other words we can " see around a corner " as every good teacher is said to do. If the sun were encircled by a ring of stars, or a nebula-like halo, the circle of light would be contracted as it passed the sun and would come to a focus at a place seventeen times the distance of Neptune, or 47,600,000,000 miles beyond the sun.

The observations made by the British expeditions during the eclipse of May 29, 1919, were not alogether satisfactory. At Principe, on account of a cloud that drifted by at an inopportune time, only a few photographs could be obtained. At Sobral one of the object glasses gave distorted images, but the other gave a very good series of seven photgraphs. These when measured at the Greenwich Observatory gave the following figures which are in accordance with those calculated by Einstein's formula.

DISPLACEMENT OF STARS IN SECONDS OF ANGLE

As observed by the British astronomers:

$$-.19-.29-.11-.20-.10-.08+.95$$

As predicted by Einstein:

$$-.22-.31-.10-.12+.04+.09+.85$$

This is regarded by the astronomers of the British Eclipse Expedition as sufficiently close to confirm Einstein's law but those who hesitate to accept so far-reaching and subversive a theory on the basis of these few minute measurements may hold their judgment in suspense until 1922 when the next solar eclipse, visible in Australia, takes place. Or possibly some means may be found to take star photographs close to the sun while shining. Our Californian mountain observatories may be of service in this since they are perched above much of the dust and mist and denser air that cause a strong light to irradiate and fog the photographic plate. Doubtless, too, the old photographs of earlier eclipses will now be got out to see if they contain any stars suitable for measuring.

Some of the opponents of Einstein suggest that the observed deflection of the starlight may be due to a

solar atmosphere that refracts the rays like our
earthly air. But it is hardly probable that an
enveloping atmosphere sufficiently dense and so far-
extending as to produce such an effect would have
remained unobserved and it is highly improbable
that the destiny of such an atmosphere should have
just the density and decrease with the distance at just
the rate to produce the deflection predicted by
Einstein's calculation.*

The discovery is rather disconcerting to astronomers, for all their calculations for the last three hundred years have been based upon the assumption
that light travels in straight lines at even speed
through empty space or, what is the same thing,
through the ether. If now light is pulled aside by
gravitation as it goes by a solid body the rays from a

---

*If you insist upon seeing just what is the difference between
Einstein's and Newton's laws of gravitation here it is as given
in *The Scientific Monthly* of January, 1920:

Any particle or light pulse moves so that the integral of ds
between the two points of its path (in four dimensions) is
stationary where

(according to Einstein)
$$ds2 = -(1-2m/r)-1dr2-r2d\theta2 + (1-2m/r)dt$$
or (according to Newton)
$$ds2 = dr2-r2d\theta2 + (1-2m/r)dt$$

These expressions are in polar coördinates for a particle of
gravitational mass m.

The new factor introduced by Einstein is, as shown above.

$$\frac{1}{1-\frac{2m}{v}}$$

distant star having to pass through the tangled throng of the Milky Way might travel a very devious route and the star would appear to us to be located in a different place from where it really is. In fact it is possible that a star which we see double may actually be single but that rays starting out from it in different directions may be so deflected by passing near other stars that when they reach us they appear to come from different points of space and so appear to us as twin stars. There may too, be dead or dark stars on the way whose existence we cannot discern and allow for.

Now those of us who are not astronomers are not much concerned over a discrepancy of a few hundredths of a second in the measurement of an angle by the telescope. We do not care much where Mercury will be five centuries hence, for we do not know quite where it is now. If astronomers made the laws of Nature instead of merely discovering them we might be afraid that at their next congress they might repeal Newton's law of gravitation and send us all flying off into space. But fortunately they have no such power and even though they should all become adherents of Einstein's most revolutionary theories, Newton's laws of mechanics and Euclid's laws of geometry would remain as true as they ever were, not perhaps absolutely and universally true, as we have assumed, but sufficiently accurate for all practical purposes. Deviations from them can only become detectible when we come to consider movements as swift as light waves or electrons.

How a heavy object can alter space relations may be seen from this simple illustration: Stretch a sheet of rubber over a hoop like a drumhead.   It is now level and flat and if parallel lines are drawn across it in two directions so as to divide it up into squares like a checkerboard all these lines are straight and equidistant and all the squares are of equal size.

A row of worms, starting in an even rank, and crawling along the parallel lines across the drumhead, would keep even all the way.   Now lay a bullet on the centre of the drumhead. The rubber sags down and stretches, most in the the middle, least at the edges.   The " parallel " lines are no longer equidistant. The squares are no longer equal. The lines are no longer of the same length. If now we repeat our worm race we shall find that those worms following lines close to the weight have to go down hill and up again and so travel a greater distance to traverse the same number of squares than those following lines nearer the edge which lie comparatively flat and are nearly as short as before. Consequently the worms will be slowed up in proportion of their nearness to the centre and the row of their heads will be swung around at an angle to their former frontage.

We might " explain " this by assuming that the worms on seeing the bullet to one side were drawn by their curiosity a little toward it, those nearest of course being drawn the most.   Or if we had got beyond this crude animistic method of explanation we might assume that the bullet was attached to the head of each worm by an invisible lariat which being pulled by the bullet drew the worms more or less to one side, the shorter the lariat the stronger the pull. Or if we had outgrown this crude mechanical method of explanation we might assume the existence of a " force " in the lead which in some mysterious manner attracts the heads of the worms inversely as the square of their distance.   But instead of inventing a wormhead psychology or an invisible cord or an incomprehensible force is it not simpler to consider the space between and to suppose that the lines to be traversed are lengthened in the neighbourhood of the weight ?

Now these four successive methods of explanation have been used to account for gravitation.   First it was assumed by the ancient Babylonians and

Hebrews that the sun and stars were living beings, gods or angels, moving of their own volition around the earth, or at least that each was guided in its orbit by its particular god or angel.   The later Greeks of Ptolemy's time supposed the heavenly bodies to be set in concentric crystal spheres and so revolved; I presume by somebody turning a crank behind the scenes.   Then came Newton and said : " Let's discard the Ptolemaic spheres and all mechanical connection and assume a force of gravitation attracting all bodies in proportion to their masses and inversely proportional to the squares of the distances separating them."   Now comes Einstein and says : " Let's discard this hypothetical force and simply assume that the field of time and space traversed by a moving body is altered if there is another body in the vicinity."   In Einstein's view gravitation is not a force; it is a distortion of space and time in the presence of matter.   A comet sweeping past the sun cannot pursue a straight course, as it could in interstellar space, but follows a curved path about the sun which is for the comet the shortest way it can go under the circumstances.

So, too, a row of light waves coming from a distant star keeps an even front as they pass through empty space, but as they come close to the sun they find their paths impeded, or, we may say, stretched. Those going nearest the sun are slowed up the most; those farthest off the least. Consequently the wave-front is slued around a bit and the direction of the ray is slightly altered.

If now light waves have difficulty getting past the sun we should expect that they would experience like difficulty getting away from the sun. They would be slowed up a bit by its gravitational pull-back. The frequency would be reduced; the interval of time between wave-crests lengthened. This means, in the case of sound, lowering the pitch. Touch your finger to the turn-table of your phonograph and you flat the tone. In the case of light, it means change of colour toward the red. This effect, according to Einstein, should be, but has not been, observed.

" If Einstein's third prediction is verified," says Sir Oliver Lodge, " Einstein's theory will dominate all higher physics and the next generation of mathematical physicists will have a terrible time of it. For university courses and for all practical purposes we shall have the Galilean and Newtonian dynamics; but they will reign as a limited monarchy and sooner or later the Einstein physics cannot fail to influence every intelligent man. If these complications are to come into science we must leave them to the younger men. I hope that gravitation, now that it has begun to interact with light, will begin to give up its secrets, but in my time I must be content to get secrets out dynamically and leave transcendental methods to others."

One English scientist, Thomas Case, writes to *The Times* to protest that it would have been in much better taste for the Royal Society to have adjourned its discussion " before bringing into question the reputation of Newton, who was President of the Royal Society for the last twenty-five years of his life and raised the society to the acme of its fame."

WHO IS EINSTEIN ?

Albert Einstein was born in Germany in 1874. He early showed the bent of his genius and at the age of twelve, when his fellow pupils were plodding along with their daily tasks, he was plunging through works of higher mathematics borrowed from his teacher. He was only eighteen when he conceived the outlines of his theory and ten years later it was ready to give to the world. He left Germany for Switzerland at the age of sixteen and became naturalized as a Swiss citizen. His first academic position was the Professorship of Mathematical Physics at the

Zürich Polytechnic. Then the founding of the Kaiser Wilhelm Academy for Research at Berlin gave him opportunity to work out his theories undisturbed by other duties. Shortly before the war he was called to Berlin to succeed the famous Dutch physicist, Professor van't Hoff in the Academy. The object of this institution was the same as Carnegie had when he founded his institution for scientific research at Washington, which was to seek out the exceptional man wherever he may be found and set him at his peculiar tasks. At Berlin Einstein receives a salary of $4,500 and has nothing to do but sit and think. This he continued to do all through the five years of war and revolution as quietly and persistently as Kant at Königsberg during the wars and revolutions of a century before. Or as Archimedes at the siege of Syracuse who was absorbed in drawing geometrical figures in the sand—his black-board—when a Roman soldier ran him through with a spear. On two occasions he took part in the world-struggle going on about his study, both actions greatly to his credit. In the beginning he refused to sign the manifesto of the German men of science denying all the charges against Germany, and at the time of the armistice he signed an appeal in favour of the revolution. He is an ardent Zionist and has promised to aid the

F

Hebrew university which is to be founded at Jerusalem.

According to tradition, Isaac Newton was led to his theory of gravitation by observing an apple falling from a tree in his garden. The newspaper correspondents start a similar tradition by reporting that Einstein got his theory of gravitation by observing a man falling from the roof of a building in Berlin. Now a man has the advantage of an apple in that he is able to tell his sensations. When Dr. Einstein, who had seen the accident from his library window in the top storey of a neighbouring apartment house, reached the spot he found the man had hit upon a pile of soft rubbish and had escaped almost without injury. Asked how it felt to fall he told Dr. Einstein that he had no sensation of a downward pull at all. This led Dr. Einstein to consider whether the relativity theory, which he had applied only to the case of uniform motion in a straight line, could not be extended to difform or accelerated motion by gravitation. So the special relativity theory which he had enunciated in 1905 developed ten years later into a generalized relativity theory *(Verallgemeinerte Relativitätstheorie)*.

HOW TO LOSE WEIGHT

A man falling out of an airplane is obeying a natural impulse, namely, the force of gravitation. So long as he does not resist he is free as air, light as a feather, and altogether comfortable. He can look down with complacency and contempt on the poor mortals below him who are trying to stand up against this natural impulse and laboriously dragging one foot after another as they crawl about the earth when they might be flying through space without effort as he is. It is only when he tries to stop his free fall by bumping against the ground that he gets into trouble on account of gravitation. It was in this way that the Calvinists, who were a sort of mathematical theologians, conceived of the fall of man. The sinner is simply obeying the force of natural depravity, namely, moral gravity, and so long as he is conscienceless and does not consider his inevitable end he has no knowledge of the moral law and is quite happy in his downfall.

A person falling freely loses all his weight. His hat does not press down on his head. His feet do not press down on his shoes. If he lets go of his walking-stick it does not " fall down " at his feet. It stands upright and simply travels along with him. For, as

Galileo showed when he dropped his big and little cannon ball off the Leaning Tower of Pisa, all bodies fall with the same speed.

If he were in a falling elevator with an opaque door he would not know he were falling unless he surmised it from the *absence of gravitation* as evidenced by his own feeling of lost weight and the queer behaviour of the objects in the car. He might fall all his life and never find it out. The law of gravitation is like criminal law; you don't feel it till you come into conflict with it.

Or if our illustration requires too tall a sky-scraper let us imagine that a comet as it flies by knocks a chip off the earth with a group of people on it. This terrestrial fragment, cast loose in space, gets caught by the attractive force of some gigantic and distant star and falls towards it with ever-increasing velocity for thousands of years. The inhabitants of this errant orb, would never know it from their own feelings or any observations they could make on *their own little world*. Does that seem incredible to you? Then tell me how do you know but that this our world is such a planet and together with the solar system has been falling for thousands of years toward some centre of attraction? Astronomers, indeed, say that we are moving at tremendous speed toward Canis Major, in other words that the world is going to the dogs.

All this means that uniformly accelerated motion, such as gravitation imparts to a freely falling body, is, like uniform translatory motion, a question of relativity and cannot be discovered by an observer carried along by such movement.

The idea that uniform translation, like the moving train we have considered, is merely relative motion, is an old idea and not hard to understand or accept. But when we try to understand the principle of relativity to acceleration, that is, to a rate of motion that is continuously increased or retarded, we get a new and revolutionary conception of the universe and are drawn into some very startling conclusions. Einstein took this step five years ago and that is what has caused the present excitement. For Einstein when he once gets hold of an idea follows it wherever it leads him with the undaunted determination of a Nantucket sailor towed by a harpooned whale. It was a whale of an idea that he harpooned in 1915 and it carried him into strange waters. It led directly to a contradiction or correction of one of the two fundamental postulates which he had laid down as the foundation of his theory of the universe in 1905, namely, that the velocity of light in space is a constant.   But he promptly abandoned this idea with cheerful nonchalance in favour of the new notion that the velocity of light is affected by gravitation.

A SUBSTITUTE FOR GRAVITY

Let us then follow Einstein and apply his Principle of Equivalence to accelerated motion and see what it leads to.   Imagine yourself shut up inside a closed chamber like an elevator car, somewhere out in space away from the gravitational forces of the earth or sun.   Suppose this chamber to be rising with a constantly increasing velocity.   We can, if we want to be definite about it, assume that the chamber is a big shell pulled up by a cable coiling around a conical windlas that hauls it up faster all the time.   Or we can assume that it is propelled from behind by the continuous backfire of explosives, like the rocket which Professor Goddard proposes to send to the moon.   All we need is some force, not gravitation, capable of giving the chamber every second an additional velocity of thirty-two feet a second.   Now the point is that if you were in such an upward-moving chamber you would not know but what you were resting on the earth.   Everything would behave exactly the same.   If you now weigh one hundred and fifty pounds on the scales, that is, if your shoe soles press down with that force, the floor of the rising chamber would press upward with that same force and you would not know the difference. If you let loose a ball from your hand the floor would rise up to meet it and it would appear to fall. If you threw the ball upward with a velocity greater than the velocity of the chamber at the moment, the ball would rise, but since the velocity of the chamber was constantly increasing the floor would gain on the ball and catch up with it. This would look to you just the same as when on earth you threw a ball into the air and it fell back to the ground, drawn, as you are accustomed to think, by " the force of gravitation." But here we have no " force," but merely a mode of motion.

Under such circumstances it would seem that all Nature conspired to keep you in the dark. You appeal to the ether, that supposedly stable and stationary medium that fills all space, but that also fails you. You try the Michelson-Morley experiment to see if you are moving through the ether or at rest on the earth but your apparatus expands or contracts just enough to deceive you.

You now try observing horizontal rays of light but they seem to bend; that is, a beam of sunshine

entering a pinhole on one side of your *camera obscura*
will not strike the wall at a spot exactly opposite but
a little below it, if you have instruments sufficiently
delicate to show this.   You try vertical rays of light
in this fashion : You examine with the spectroscope
rays of light coming from two sources below (behind)
your instrument, one at a distance and the other
nearer.   Now since you are moving away with increasing speed, the light from the farther source will
have to take longer strides to catch up.   Or in other
words, its frequency will be reduced and it will be
shoved toward the red end of the spectrum where
the longer waves are.   You will have noticed that
when a whistling train rushes past the train you are
on, the whistle as it comes towards you is raised in
pitch (decreased wave-length) and as it recedes from
you is lowered in pitch (increased wave-length).

Now, says Einstein to himself, if my Principle of
Equivalence is correct and there is no difference between (1) the weight and (2) the accelerated upward
movement of an observer, then all the optical effects
that I have thought out in the second case must apply
to the first, that is, to gravitation.   It must follow
that a ray of light passing through a gravitational
field will be bent out of its course as though it were
attracted by the heavy body.   This prediction has

been verified. It must further follow that light proceeding from a heavy body like the sun or a star will be held back or slowed up by the attraction of gravitation, and the spectral lines will be displaced toward the left as compared with the same lines in the spectrum of an earthly light. Now such displacement has been observed in stellar spectra but it does not seem to be of the right value to satisfy Einstein's equation and it has not been observed in sunlight.

The remarkable thing about it is that Einstein, by following a line of reasoning somewhat like that which I have crudely outlined, not merely supplied an explanation for phenomena that had been observed but could not be explained (such as the discrepancy in the orbit of Mercury) but he provided in advance an explanation for phenomena that had never been observed until he directed attention to it (such as the deflection of starlight by the sun). Sir Oliver Lodge says of this :*

Before Einstein's prediction nothing of the kind had been seen, nothing of the kind had been looked for, nor, so far as it is known, had such an amount of deflection been suspected.

*Nineteenth Century. December 1919

Whatever may ultimately be thought of the validity of Einstein's views as a whole it is evident that he has worked out a mathematical method of unprecedented power and wide usefulness.

Professor Bumstead of Yale says:

> Einstein's theory is important in that it exemplifies a method which is in many respects new in theoretical physics and which may prove to be a very powerful method for advancing scientific knowledge. There was no idea that the prediction of the bending of light would fix up Mercury's perihelion and incidentally explain a two-century old astronomical difficulty. That came straight out of a blue sky.

MECHANICAL VERSUS MATHEMATICAL MINDS

We sometimes hear it said that "Einstein has overthrown Newton's theory of gravitation." That is impossible because Newton did not have any theory of gravitation. He merely laid down the law of gravitation. He told how bodies behaved toward their neighbours; he did not tell why. Newton was not content with the idea of action at a distance through empty space and he tried to explain gravitation by the pressure of the ether on material bodies but he was not satisfied with the results and did not publish them. In the 234 years since many men have tried their hands at devising some sort of machinery

that will " explain " gravitation. For human beings
are like Toddie of " Helen's Babies " and want to
have the watch opened so they can " see the wheels
go wound." At least Anglo-Saxons have that desire.
Poincaré, the French physicist, said this is the distinction between the Anglo-Saxon and Latin minds;
the former are uneasy until they can imagine a
mechanical model to represent natural phenomena,
the latter are satisfied with a mathematical
formula expressing the action. The ether,
which was invented to explain light, also required
" explanation." Lord Kelvin imagined it to consist
of spinning tops which have a sort of mobile stability.
Sir Oliver Lodge has filled it with a complicated
structure of interlocking geared wheels to account for
electromagnetic action. These are typical Anglo-
Saxon modes of thinking. On the other hand, Einstein, who, in spite of his Hebrew blood and German
training, has preëminently what Poincaré claims is
the Latin temperament, does not have any use for
the ether and does not care at all whether he can
" picture " the fourth dimensions on paper or not.

Now some of us are excessively Anglo-Saxon in our
attitude towards mathematics. It is with a fellow-
feeling for such folks that I have filled this little
volume with such crude and absurd analogies as

trains and elevators, and projectiles flying through space and Coney Island mirrors. To the mathematically minded such illusrations are not simplifications but complications, not representations but caricatures. Mathematics is the proper language of physics as the five-barred staff is the proper language of music. Ask a musician to explain a symphony in plain everyday English and he cannot do it, though he carry the Oxford Dictionary in his head. He can have the music played for us or he can show us the printed score but he could never convey it in ordinary language however long he might be willing to talk or we to listen. But we must not do the musician or the mathematician the injustice to suspect that his notions are hazy or absurd because he cannot explain (*i.e.* translate) them to us.

Nor should we assume that the new ideas, because they are more difficult for us to grasp, are necessarily more complicated or extravagant than the old. A friend of mine who is familiar with both tells me that Einstein's papers are easier reading than Newton's " Principia."

The aim of science is simplification through generalization and this is the widest generalization yet attempted. It promises to bring gravitation into relationship with other forces. One great generalization, the law of the conservation of energy worked out by Joule and others in the forties, brought heat and work and chemical power all into one simple system. Clerk Maxwell in the seventies brought together in one beautiful formulation all the diverse phenomena of light, electricity and magnetism.

But gravitation has always stood out against any such league of natural forces. It refused to come into the combine. It remained unique, independent, irreducible, unalterable and inexplicable. Everything else is correlated and interactive. Heat destroys magnetism, magnetism produces electricity; electricity dissolves chemical combination; chemical combination produces heat; heat causes motion; motion makes magnetism; magnetism produces heat; and so on in endless round, each affecting all the others. Different substances behave very differently; one is more easily heated than another; some are readily magnetized or electrified, others are not so susceptible; certain elements rush into each other's arms, others cannot be forced into combination.

But gravitation seemed indifferent to all these things; it showed no prejudices or preferences. It attracted with equal force all sorts of substances, no matter whether they were hot or cold, shiny or black, moving or still, electrified or magnetized or neither.

Other forces and effects too required time for action at a distance. Sound travels at the rate of 1,100 feet a second in ordinary air. Light travels at the rate of 186,337 miles a second in a vacuum. But the force of gravity seemed not to require any time but to be everywhere, acting all the while, and nothing could shield it off or shut it out or in any way interfere with it. The substance or mass of a body as measured by its weight (the gravitational pull of the earth) was always identical with its mass as measured by its inertia (its resistance to being set in motion). All the energies are interchangeable. All other forces could be reduced or increased, annulled or brought into effect at will. Not so gravitation. Any bodies of a certain mass placed at a certain distance apart are always drawn by the same attraction. That is, gravitation is affected by nothing except geometrical relationships.

This naturally leads us to suspect that gravitation is nothing but a geometrical relationship, that it is somehow a peculiarity of space itself. If so, our demand of the physicist that he show us gravitation— drag out this mysterious force from its hiding-place and let us see it—is altogether irrational. It is like a blind man hunting in a dark cellar at midnight for a black cat that isn't there. The geometrician tells us that the internal angles of any triangle are equal to two right angles. Shall we ask him, what is the force that makes it so? Shall we refuse to ride on a trolley car until the electrician can answer our persistent question; "but what is electricity?" When we ask such a question we are really asking him to tell us what electricity is *not*. To show us what electricity is he can keep his mouth shut and simply point the dynamo that produces it, the wire that conveys it and the motor that consumes it. But what we secretly mean is that he shows us a mechanical model that imperfectly imitates some of the actions of electricity or a mathematical formula that will calculate its effects.

Now Einstein seems in the way of making gravitation the foundation of a new system of geometry. Instead of "explaining" gravitation in terms of something else he will explain other things in terms of gravitation, or rather of his space-time manifold of which gravitation is one of the properties.

Einstein's *law* of gravitation proves to be more accurate than Newton's law, but the correction is trifling except in rare cases. But Einsein's *theory* of gravitation is fundamental and far-reaching and if it is substantiated it will revolutionize physics and radically affect our ordinary conceptions of the

universe.   The verification of a prediction does not
necessarily prove the truth of the hypothesis that led
to the prediction.   Many a scientific discovery has
come out of a false assumption.   Just as a miner may
reopen an abandoned gold mine or work over his
dump heap to get more out of it, so scientists often
return to an old theory which they had abandoned for
a more fruitful hypothesis.

THE WEIGHT OF LIGHT

It is interesting to see that our modern physicists
show a disposition to adopt a corpuscular or emission
theory of light not unlike the conception which
Newton steadfastly and vainly defended against the
undulatory theory.   Professor Thomson, of Cambridge, reminds us that the crucial experiment
between the two theories was the test made by
Bennet in 1792 to determine if light exerted any
pressure on a body when it struck it as it would if
light consisted of minute particles driven straight
forward with great velocity.   Bennet found no such
pressure and the corpuscular theory was regarded as
disproved.   But it was later found that the
undulatory theory also involved such a pressure, and
recent experimenters have proved and measured it.
As Professor Thomson says :

It is perhaps fortunate that Bennet had not at his command more delicate apparatus. Had he discovered the pressure of light, it would have shaken confidence in the undulatory theory and checked that magnificent work at the beginning of the last century which so greatly increased our knowledge of optics.

Of course any modern form of the emission theory of light must account, as Newton's did not, for such phenomena as interference and polarization, which are so satisfactorily handled by the undulatory theory. That is, it must combine the best features of both. Profesor Thomson shows that only an exceedingly small fraction of the ether is concerned in the forward movement of light, in other words, "the wave front must be more analogous to bright specks on a dark ground than to a uniformly illuminated surface." He does not, however, go so far as Planck in regarding it as proved that radiant energy of all kinds has a unit or atomic structure, the colour of the light depending on the size of these particles.

The discovery of the pressure of a beam of light has led to some startling conclusions. For example, what shall be done with Newton's law that action and reaction are equal? When a gun is fired the kick of the gun is balanced by the momentum of the projectile. When a reflector throws a beam of light into space, the kick of it is there all right but where is the projectile, if light is merely the undulation of

G

an imponderable fluid?   We may suppose that the
light strikes some dark body out in space, transmits
its impulse to that and Newton's law is satisfied,
but it may be a long time before such a body is en-
countered and it may never be : at any rate a law
that remains in a state of innocuous desuetude for
several thousand years is not good for much.   We
must then assume that light has mass since it has
inertia and momentum. But if light has mass it
must have weight; that is, it must be attracted by
gravitation. The eclipse observations confirmed this
deduction. Newton would have expected something
of this, for he says in his *Opticks*.

Query 1.—Do not Bodies act upon Light at a distance, and by
their action bend its Rays, and is not this action *(caeteris paribus)*
strongest at the least distance?*

The observed deflection of light due to the sun's
gravitation is greater than Newton would have
anticipated but it would have been still more disconcerting to the nineteenth-century physicists, for in
giving up Newton's emission theory they had come
to regard light as merely a form of motion in a
weightless medium, the ether. Disembodied energy,
like heat and light in ethereal space, was regarded as

*Quoted by   Eddington in   *Contemporary Review,*
December, 1919.

having no mass or weight.   Twentieth-century physicists are coming to the opposite view, that the mass of a body is the measure of its internal energy. If so, mass is not constant but changes with composition, temperature, structure, electrification and motion.

As Einstein himself expresses it :

It is evident that it is not possible to attribute an absolute sense to the notion of acceleration, no more than to the notion of velocity. It is only possible to speak of the acceleration of a material point in connection with a body taken as the body of reference. It follows from this that there is no sense in attributing to a body a " resistance to acceleration " in the absolute sense like the resistance of inertia in the classical mechanics. Further, this resistance of inertia ought to be so much the greater when there is, in the neighbourhood of the body, more inert masses not in accelerated movement. On the other hand, this resistance ought to disappear when these masses participate in the acceleration of the body.

Now it is altogether remarkable that the equations of the gravitational field contain these different aspects of the resistance of inertia, which one might call the *relativity of inertia.*

The progress of science is continually toward a dematerialization of matter. Physical analysis resolves the crude, heavy, solid stuff that our senses show us into finer and finer particles further and further apart until these practically disappear into mere points of irradiating influence. First the mass is divided into the molecule and this again into the atom, assumed, at the time it was invented, to be the ultimate unit

of matter. But recently the atom has been shown to
be a sort of solar system, but more complex, com-
posed of hundreds of electrons, corpuscles of electricity, whose radius is calculated to be 1/10,000,000,
000,000 of a centimetre (a centimetre is so —— long).
" But the size of the centres of disturbance, which in
Einstein's theory are associated with matter, bears
to the size of the electron about the same proportion
as the size of the smallest particle visible under the
most powerful microscope to that of the earth
itself."*

The old axiom was "matter cannot act where it is
not." The new version might rather read : " matter
cannot act except where it is not."   That is to say,
attention is now directed to the space surrounding a
material body or electrical corpuscle.

Although we laymen are not concerned with the
niceties of astronomical measurements there is an
aspect of this conflict of theories that does interest
us. The theory of Newton or, to go back further, of
Galileo, that the earth moves around the sun, altered
profoundly the philosophical and religious beliefs of
the world, and the theory of Einstein is much more
far-reaching and revolutionary in its metaphysical

*Sir Joseph Thomson in *Nature*, December 4, 1919.

implications than the former. Professor Planck, who has just received the Nobel Prize for his discoveries in physics, said of Einstein's first paper:

It surpasses in boldness everything previously suggested in speculative natural philosophy and even in the philosophical theories of knowledge. Non-Euclidean geometry is child's play in comparison . . . The revolution introduced into the physical conceptions of the world is only to be compared in extent and depth with that brought about by the Copernican system of the universe.

MUTABLE THEORIES AND STABLE FACTS

There is a feeling very prevalent among the general public interested in such things that the foundations of modern science are being swept away by the recent discoveries. The layman has been led to believe that such laws as gravitation, the conservation of matter and the immutability of the elements are the most certain and absolute truths of science. But now he hears reputable men of science talk calmly about the decay of matter and the transformation of one element into another, and gravely consider a theory which makes invalid Newton's three laws of motion. It surprises, even shocks him, as much as it would to have a convention of bishops discuss the question of whether there is a God, or the Supreme Court agree to set aside the Constitution of the United States, or a congress of physicians resolve that all medicine does more harm than good.   He knows that the mere broaching of such heretical views in these assemblies would be met with a storm of indignation and that all the weapons of contempt, ridicule and even personal spite would be directed against the rash innovator.   Therefore he is astonished and puzzled to see that in the scientific world these revolutionary theories are received with interest and even pleasure, and in the criticism to which they are subjected there is scarcely a trace of animosity.   And he does not see why men of science who have accepted doctrines apparently contradictory to their former teachings do not appear shamefaced and apologetic before the public, like augurs whose tricks had been exposed.

The difficulty of the layman arises from his not understanding how a scientist looks at his science; not realizing how firmly he holds to its facts and how loosely he holds to its theories.   The scientist never bothers his head with the question whether a particular theory is true or false.   He considers it simply as more or less useful, more or less adequate, succinct and comprehensive.   A theory is merely a tool, and he drops one theory and picks up another at will and without a thought of inconsistency, just as a carpenter drops his saw and picks up his chisel.   He will say that the earth moves around the sun one moment and the next will revert to the theory of Chaldean astronomers, because it is more convenient, and say " the sun rises."

Really, the new discoveries are not so upsetting to science as they appear to the general public. Unexpected and revolutionary as they are, no page of millions that record the experiments and observations of science is invalidated. No man's work is proved wrong. Revolutions in science do not destroy : they extend.

In the reaction of public opinion toward any novel and revolutionary idea there are three stages observable.

1. That it is not true.

2. That it is not new even if it is true.

3. That it does not make any difference anyhow.

The first is merely the natural and instinctive reaction against any disturbing intellectual innovation. It is a flat denial inspired by that unconscious neophobia or xenophobia that possesses all of us more or less. The second stage is the effort at compromise in which usually both the advocates and opponents of the new idea coöperate by endeavouring to prove that it is not so novel and unprecedented as was at first assumed but fits in very fairly with our accepted notions, in fact may be regarded as a supplement or even a natural development of them.   The third stage, like the second, is designed as an attempt to disarm opposition by allaying alarm in the conservative mind.

The second line of argument has a good deal of validity, for even the most startling and original idea will be found on closer examination to have its roots deep in the ground of the past and to have been approximately anticipated many times before.   The third line of argument also contains some truth for we find everyday life does go on in much the same way, although it may seem that the foundations have been knocked from under our mental, moral or social universe by some new notion.   Yet as the popular mind gradually accepts and adapts itself to the novel conception we generally find that its influence is even more far-reaching than was at first anticipated.

In the case of the Copernican theory it took about two centuries for the controversy to pass through the three stages and the mind of the public to become readjusted to the new conception of the earth's revolution.   In the case of the Darwinian theory of evolution the process was accomplished in about fifty years.   The Einstein theory is more subversive of ordinary ideas than either of the others so it would naturally take longer to soak in.   But the modern mind seems to be subject to acceleration and we see in the two months since the notion has been sprung upon the public that all three of the lines of argument are appearng at once and so the controversial period may run its course in five years though it will be longer before its indirect influence upon our fundamental philosophy and habits of thought are fully felt.

SCIENTIFIC VERSUS LEGAL LAWS

In all such discussions we must bear in mind that " law" in the scientific sense of the word means, not a commandment or a rule, but merely a way of working.   It is a concise description of how things behave.   There are no laws in Nature; there are only laws of Nature, that is to say, laws drawn out of Nature (or, if you prefer Latin to Anglo-Saxon, laws deduced from Nature) by man for his own convenience in thinking.   Physical laws are therefore essentially phychological : mere memory schemes, calculating machines.   The law of gravitation is no more gravity than the funny wriggles that my stenographer is making in her notebook are the sounds I am uttering.   To change geometries does not require any

such effort as to change cars. It means merely
changing our minds. But this is harder for some of
us than it ought to be. Here is where the theory of
relativity will be of use to us. Poincaré, the French
mathematician, cousin of the late President, said:
"These two propositions, ' the earth turns round,'
and 'it is more convenient to suppose the earth turns
round,' have the same meaning. There is nothing
more in the one than in the other." If Galileo and
his inquisitors had understood the principle of Relativity it might have saved them both trouble; the
former temporary imprisonment and the latter ever-
lasting disgrace. A revolution in science is simply a
change in mental attitude. Maybe a political revolution is no more.

It is disconcerting to the layman to be told, first
that matter consists of solid round atoms in empty
space; next, that it is made of mere particles of electricity and negative at that; then that it is consti-
tuted out of strains in the ether; again, that the
atoms are bubbles in the ether; and finally, that there
is not any ether. But these various hypotheses are
like the crayon strokes that an artist makes about a
figure he is trying to draw. They are all attempts
at preliminary sketches for mental pictures of natural
phenomena. We do not call the geographers incon-

sistent and contradictory because one colours Massachusetts red on the map and another colours it green. All scientific hypotheses are put to the pragmatic test of which works the best in unlocking the secrets of nature. Is " wheat " or " sesame " the magic word? Whether we call a dog " Fido " or " Towser " depends not on which name is shorter or sounds better, but on which the dog answers to. If gravitation comes to heel better when we say " Einstein " than when we say " Newton," all right, we'll change. I trust that these frivolous illustrations will not lead my readers to accuse me of treating gravity with levity.

The layman—and with him must be included all those who have merely learned science but not used it—talks a great deal about " the laws of Nature," which he regards as abstract, immutable, universal and eternal edicts, part of which are transcribed into the text-books. To the working scientist they are only more or less convenient formulas; in the ultimate analysis only mnemonic symbols for stringing together facts to make them easier to handle, like *vibgyor*, for the spec um colours. He knows that most of them are limited in their scope and only approximate in their accuracy. His chief delight is in discovering these limitations and irregularities. He regards these "laws" with no awe or reverence. He has no attachment for any of them—unless it happens to be one that he has formulated himself. If he finds a new hypothesis that works better he throws the old one aside as he does his old model dynamo, or keeps it around as handy still for doing some of the common work of the laboratory. It is, to recur to our example, just as "true," using the word in its ordinary sense, to say that the sun goes around the earth as to say that the earth goes around the sun, for all motion is relative, and we can regard either body as the stationary one or both as moving, as we choose. When we say that the statement that the earth moves around the sun is the "true" one, we merely mean that it is the more convenient form of expression, for on this hypothesis the paths of the earth and the other planets become circles (or more accurately speaking, irregular and eccentric spirals) while on the other and older hypothesis their paths are very complicated and difficult to handle mathematically. The theory that the earth moves is not only simpler than that of a stationary earth, but it is wider in its scope. It explains more, that is, it connects up with other knowledge, such as the flattening at the poles. Copernicus, then, did not discover a new fact about the solar system. He only invented a lazier way of thinking about it.

The man of science invents an hypothesis whenever he needs one in his business. It is to him merely a new tool, a *novum organum*. If there is not an ether it would be necessary to create one. So he did it. He had to have a noun for the verb "undulate." When he had created it he saw it was not good. The properties with which he endowed it were self-contradictory, and it refused either to move with the earth or to pass through it. But these theoretical inconsistencies do not bother the physicist much. In spite of them the ether is a handy thing to have about the laboratory. The scientist does not abandon a theory because it has inconsistencies any more than he divorces his wife because she has inconsistencies. Certainly the physicist did not consider himself presumptuous in thus inventing ether for his own convenience. He knew that the ordinary man had in the same way invented "matter" long ago for his own convenience. It is a crude, inadequate and impossible idea, this naïve conception of matter as something solid, heavy, hard, inert, indestructible, impenetrable, coloured and surfaced; but it is good enough for part of the people all of the time and for all of the people part of the

time.   The physicist himself uses it for everyday.
Only in his rigorous moments does he come down to
bed-rock and say, with Poincaré, "Mass is a co-
efficient which it is convenient to introduce into
calculations."

But when the physicist thus reduces matter to a
small italic m some people are sure to say that he is
denying the existence of matter.   What would they
say about Riemann who considers matter to be holes
in the ether? A definition is a different thing from a
denial.   There are people among us who deny the
existence of matter and they call themselves
" Scientists," too, but they are not the ones who are
devoting their days and nights to the study of the
workings of matter in order to make it the servant of
man.

A professor of chemistry would not think of asking
his students if the atomic theory is true any more
than he would ask them if the atomic theory is blue.
He does not care whether they believe the atomic
theory or not. He only wants them to be able to use
the atomic theory for getting certain valuable results.
Consequently, he watches with interest and without
apprehension the progress of discovery in radio-
activity which is undermining the old conception of
the atom. He would be glad to get rid of the atomic

theory if he could find something better because after all it is a clumsy thing and will not hold half the facts he wants to put into it. He would have no more hesitation about dropping it than he has in setting down one beaker to pick up a larger one when what he has in the first is frothing over. He does not want to spill anything, but he does not care what vessel it is in. Revolutions in science never go backward and they differ from political revolutions in that nothing worth saving is lost in transition. The new theory must always include all that the old one does and more. In their struggle for existence, formulas fight like snakes; the one that can swallow the other beats. Now a four-dimensional universe can take in a three dimensional universe and have space to spare for whatever the narrower conception could not include so it seems likely to prevail.

We now know how to sympathize with those poor frightened people who lived in the times of Copernicus and Galileo when they were told that the solid earth on which they stood was not supported by anything, but whirling about and rushing around through empty space and that half the time they hung with their heads down over immeasurable space with nothing to hold on to. But they got used to it in time and lived happily ever after. So may we.

For the benefit of those who want to get their information at first hand I append an article by Dr. Einstein himself which appeared in the *Times* of December 13, 1918 (for permission to re-print which the publishers are indebted to the Editor), and in *Science* of January 6, 1920.

TIME, SPACE, AND GRAVITATION

By Dr. Albert Einstein

I respond with pleasure to your Correspondent's request that I should write something for the *Times* on the Theory of Relativity.

After the lamentable breach in the former internaformal relations, existing among men of science, it is with joy and gratefulness that I accept this opportunity of communications with English astronomers and physicists. It was in accordance with the high and proud tradition of English science that English scientific men should have given their time and labour, and that English institutions should have provided the material means, to test a theory that had been completed and published in the country of their enemies in the midst of war. Although investigation of the influence of the solar gravitational field on rays of light is a purely objective matter, I am none the less very glad to express my personal thanks to my English colleagues in this branch of science; for without their aid I should not have obtained proof of the most vital deduction from my theory.

There are several kinds of theory in Physics. Most of them are constructive. These attempt to build a picture of complex phenomena out of some relatively simple proposition. The kinetic theory of gases, for instance, attempts to refer to molecular movement the mechanical, thermal, and diffusional properties of gases. When we say that we understand a group of natural phenomena, we mean that we have found a constructive theory which embraces them.

But in addition to this most weighty group of theories, there is another group consisting of what I call theories of principle. These employ the analytic, not the synthetic method. Their starting-point and foundation are not hypothetical constituents, but empirically observed general properties of phenomena, principles from which mathematical formulæ are deduced of such a kind that they apply to every case which presents itself. Thermodynamics, for instance, starting from the fact that perpetual motion never occurs in ordinary experience, attempts to deduce from this, by analytic processes, a theory which will apply in every case. The merit of constructive theories is their comprehensiveness,

H

adaptability, and clarity, that of the theories of principle, their logical perfection, and the security of their foundation.

The theory of relativity is a theory of principle. To understand it, the principles on which it rests must be grasped.   But before starting these it is necessary to point out that the theory of relativity is like a house with two separate storeys, the special relativity theory and the general theory of relativity.

Since the time of the ancient Greeks it has been well known that in describing the motion of a body we must refer to another body.   The motion of a railway train is described with reference to the ground, of a planet with reference to the total assemblage of visible fixed stars.   In physics the bodies to which motions are spatially referred are termed systems of coördinates.   The laws of mechanics of Galileo and Newton can be formulated only by using a system of coördinates.

The state of motion of a system of coördinates cannot be chosen arbitrarily if the laws of mechanics are to hold good (it must be free from twisting and from acceleration).   The system of coördinates employed in mechanics is called an inertia-system.   The state of motion of an inertia-system, so far as mechanics are concerned, is not restricted by nature

to one condition. The condition in the following proposition suffices : a system of coördinates moving in
the same direction and at the same rate as a system
of inertia is itself a system of inertia. The special
relativity theory is therefore the application of the
following proposition to any natural process :—
" Every law of nature which holds good with respect
to a coördinate system K must also hold good for any
other system K', provided that K and K' are in uniform movement of translation."

The second principle on which the special relativity
theory rests is that of the constancy of the velocity of
light in a vacuum. Light in a vacuum has a definite
and constant velocity, independent of the velocity of
its source. Physicists owe their confidence in this
proposition to the Maxwell-Lorentz theory of electro-
dynamics.

The two principles which I have mentioned have
received strong experimental confirmation, but do
not seem to be logically compatible. The special
relativity theory achieved their logical reconciliation
by making a change in kinematics, that is to say, in
the doctrine of the physical laws of space and time.
It became evident that a statement of the coincidence
of two events could have a meaning only in connection with a system of coördinates, that the mass of

bodies and the rate of movement of clocks must depend on their state of motion with regard to the coördinates.

But the older physics, including the laws of motion of Galileo and Newton, clashed with the relativistic kinematics that I have indicated. The latter gave origin to certain generalized mathematical conditions with which the laws of nature would have to conform if the two fundamental principles were compatible. Physics had to be modified. The most notable change was a new law of motion for (very rapidly) moving mass-points, and this soon came to be verified in the case of electrically-laden particles. The most important result of the special relativity system concerned the inert mass of a material system. It became evident that the inertia of such a system must depend on its energy-content, so that we were driven to the conception that inert mass was nothing else than latent energy. The doctrine of the conservation of mass lost its independence and became merged in the doctrine of conservation of energy.

The special relativity theory, which was simply a systematic extension of the electro-dynamics of Maxwell and Lorentz, had consequences which reached beyond itself. Must the independence of physical laws with regard to a system of coördinates be limited to systems of coördinates in uniform movement of translation with regard to one another? What has nature to do with the coördinate systems that we propose and with their motions? Although it may be necessary for our descriptions of nature to employ systems of coördinates that we have selected arbitrarily, the choice should not be limited in any way so far as their state of motion is concerned. (General theory of relativity.) The application of this general theory of relativity was found to be in conflict with a well-known experiment, according to which it appeared that the weight and the inertia of a body depended on the same constants (identity of inert and heavy masses). Consider the case of a system of coördinates which is conceived as being in stable rotation relative to a system of inertia in the Newtonian sense. The forces which, relatively to this system, are centrifugal must, in the Newtonian sense, be attributed to inertia. But these centrifugal forces are, like gravitation, proportional to the mass of the bodies. Is it not, then, possible to regard the system of coördinates as at rest, and the centrifugal forces as gravitational? The interpretation seems obvious, but classical mechanics forbade it.

This slight sketch indicates how a generalized theory of relativity must include the laws of gravitation, and actual pursuit of the conception has justified the hope. But the way was harder than was expected, because it contracted Euclidean geometry. In other words, the laws according to which material bodies are arranged in space do not exactly agree with the laws of space prescribed by the Euclidean geometry of solids. This is what is meant by the phrase "a warp in space." The fundamental concepts " straight," " plane," etc., accordingly lose their exact meaning in physics.

In the generalized theory of relativity, the doctrine of space and time, kinematics, is no longer one of the absolute foundations of general physics. The geometrical states of bodies and the rates of clocks depend in the first place on their gravitational fields, which again are produced by the material systems concerned.

Thus the new theory of gravitation diverges widely from that of Newton with respect to its basal principle. But in practical application the two agree so closely that it has been difficult to find cases in which the actual differences could be subjected to observation. As yet only the following have been suggested :—

1. The distortion of the oval orbits of planets

round the sun (confirmed in the case of the planet Mercury).

2. The deviation of light-rays in a gravitational field (confirmed by the English Solar Eclipse expedition).

3. The shifting of spectral lines towards the red end of the spectrum in the case of light coming to us from stars of appreciable mass (not yet confirmed).

The great attraction of the theory is its logical consistency. If any deduction from it should prove untenable, it must be given up. A modification of it seems impossible without destruction of the whole.

No one must think that Newton's great creation can be overthrown in any real sense by this or by any other theory. His clear and wide ideas will forever retain their significance as the foundation on which our modern conceptions of physics have been built.

A final comment. The description of me and my circumstances in *The Times* shows an amusing feat of imagination on the part of the writer. By an application of the theory of relativity to the taste of readers, to-day in Germany I am called a German man of science, and in England I am represented as a Swiss Jew. If I come to be regarded as a *bête noire,* the descriptions will be reversed, and I shall become a Swiss Jew for the Germans and a German man of science for the English.

And finally

IF YOU WANT TO READ MORE ABOUT THE EINSTEIN THEORIES

For the non-mathematical reader:

ABBOT, EDWIN.
    Flatland, by A Square. Boston, 1891.
    An amusing way of leading up to the fourth dimension.
CAMPBELL, NORMAN.
    The Commonsense of Relativity. *Philosophical Magazine*, April, 1911.
CARR, WILDON.
    The Metaphysical Implications of the Theory of Relativity. *Philosophical Review*, Jan., 1915.
CARUS, PAUL.
    The Principle of Relativity. Chicago: Open Court Publishing Co., 1913.
COMSTOCK, D. F.
    The Principle of Relativity. *Science*, May 20, 1910, vol. 31, p. 767.
CUNNINGHAM, E.
    Einstein's Relativity Theory of Gravitation. *Nature*, Dec. 4. 11, and 18, 1919.
    An interesting non-mathematical discussion of the latest phases of the theory.
EDDINGTON, A. S.
    Einstein's Theory of Space and Time. *Contemporary Review*, Dec. 1919.
    Good popular article.
EDDINGTON, A. S.
    Gravitation. *Scientific American Supplement*, July 6 and 13, 1918.
    An excellent popular explanation by the leading British disciple of Enstein.
EINSTEIN, A.
    Time, Space, and Gravitation. *Science*, Garrison, 1920 Jan. 3., n.s., vol. 51, p. 8-10.
    My Theory. *Living Age*. Boston, 1920, vol. 304. p. 41-3, Jan 3.
FLAMMARION, CAMILLE.
    Lumen. New York: Dodd, Mead and Co., 1897.
    Contains nothing about Einsten but presents the relativity of time in fantastic form.
KEYSER, C. J.
    Concerning the Figure and the Dimensions of the Universe of Space. *Science*, June 13, 1913.
LODGE, SIR OLIVER.

The New Theory of Gravity. *Nineteenth Century*, Dec., 1919.

The Ether versus Relativity. *Fortnightly Review*, Jan., 1920.

Admirable article by a courteous opponent.

POINCARÉ, HENRI.

Science and Method; also contained in The Foundations of Science. New York: The Science Press. 1913.

RUSSELL, BERTRAND.

The Relativity Theory of Gravitation. *English Review*, Dec., 1919.

A clear explanation by one of the foremost of British philosophers.

THOMSON, J.

Deflection of Light by Gravitation and the Einstein Theory of Relativity. *Scientific Monthly*, Garrison, N. Y., 1920, vol. 10, p. 79-85, Jan.

VARIOUS WRITERS.

The Fourth Dimension Simply Explained. New York Munn and Co., 1910.

The essays submitted for a prize offered by the *Scientific American*. Twenty-two mathematicians try their best to justify the title and if they do not succeed it is not their fault.

WETZEL, REINHARD A.

The New Relativity in Physics. *Science*, New York, 1913. New ser., vol. 38, pp. 466-474.

Explains the relativity of time with diagrams and references to the literature.

Other articles of general interest may be found in:

Science: July 16, 1909; May 20, 1910; June 20, 1913; April 24, 1914; Dec. 5, 1919.

Scientific American Supplement: April 7, 1917: Dec. 17, 1910.

London *Nation:* Nov. 14 and Dec. 27, 1919.

London *Times:* Nov. 8, 18, and 25, Dec. 4 and 19, 1919.

London *Nature:* Almost every number in Nov., Dec., 1919, Jan., Feb., 1920.

New York *Times:* Nov. 7 and 16, Dec. 21, 1919.

New York *Sun:* Nov. 10, 1919.

For the mathematical reader:

EINSTEIN, ALBERT.

Bases physiques d'une théorie de la gravitation. Société Astronomique de France. *Bulletin*, Paris, 1917, Tome. 31, pp. 407-411.

Die formale Grundlage der allgemeinen Relativitäts-theorie. Köoniglich Preussischen Akademie der Wissen-

WORKS ON EINSTEIN THEORY 123

schaften. *Sitsungsberichte*, Berlin, 1904 (Juli-Dez), pp. 1030-1085.

Die Grundlagen der allgemeinen. Relativitätstheorie. *Annalen der Physik*, Leipzig, 1916, Band 49, Folge 4, p. 769.

Ist die Trägheit eines Korpers von seinem Energieinhalt abhängig? *Annalen der Physik*, Leipzig, 1905, Band 18, Folge 4, pp. 639-644.

First attempt at applying his theory of gravitation to the internal structure of the atom.

Lichtgeschwindigkeit und Statik des Gravitätionsfeldes *Annelan der Physik*, Leipzig, 1912, Band 38, Folge 4, pp. 355-369.

Prinzipielles zur allgemeinen Relativitätstheorie. *Annalen der Physik*, Leipzig, 1918, Band 55, Folge, 4 pp. 241-244.

Uber das Relativitätsprinzip und die aus demselben gezogenen Folgerungen. *Jahrouch der Radioaktivität und Elektronik*, Lipzig, 1908, Band 4, pp. 411-462.

Uber den Einfluss der Schwerkraft auf die Ausbreitung des Lichtes. *Annalen der Physik*, Lypzig, 1911, Band 35, p. 898-908.

Uber die Möglichkeit einer neuen Prüfung des Relativitätsprinzips. *Annalen der Physik*, Leipzig, 1907, Band 23 p. 197-208.

Uber die vom Relativitätsprinzip geforderte Trägheit der Energie. *Annalen der Physik*. Leipzig, 1907. Band 23, Folge 4, pp. 371-384.

Uber einen die Erzeugung und Verwandlung des Lichtes betreffendes heuristischen Gesichtspunkt. *Annalen der Physik*, Leipzig, 1905, Band 17, Folge 4, pp. 132-148.

Zum gegenwärtigen Stande des Gravitätionsproblems *Physikalische Zeitschrift*, Leipzig, 1913, Band 14, pp. 1249-1266.

Zum Relativitäts Problem. *Scientia, Bolgna*, 1914, vol 15, pp. 337-48.

Zur Elektrodynamik bewegter Korper. *Annalen der Physik*, Leipzig, 1905, Band 17, Folge 4. pp. 891-921.

Zur Theorie der Lichterzeugung und Lichtabsorption *Annalen der Physik*, Leipzig, 1906. Band 20, Folge 4, pp. 199-206.

Spielen Gravitationsfelder im Aufbau der materiellen Elementarteilchen eine wesentliche Rolle? *Sitz. Preuss. Akad. Wiss.*, April 10, 1919.

EINSTEIN, ALBERT, and MARCEL GROSSMAN.
Entwurf einer verallgemeinerten Relativitätstheorie und einer Theorie der Gravitation. *Zeitschrift für Mathematik und Pyhsik*, Leipzig, 1914, Band 62, pp. 225-261.

124 WORKS ON EINSTEIN THEORY

EINSTEIN, LORENTZ and MINKOWSKI.
> Relativitätsprinzip, Das. Eine Sammlung von Abhand-
> lungen. Mit Anmerkungen von A. Sommerfeld und Vor-
> wort von O. Blumenthal. Leipzig: B. G. Teubner, 1913.
> 89 p. *(Fortschritte der mathematichen Wissenschaften in
> Monographien, Heft, 2.)*
> A reprint of these three fundamental papers.

ABRAHAM, M.
> Die neue Mechanik. *Scientia*, Bolgna, 1914, vol. 15,
> pp. 8-27.
> Nochmals Relativät und Gravitation Bemerkungen zu A.
> Einsteins Erwiderung. *Annalen der Physik*, Leipzig, 1912,
> Folge 4, vol. 39, pp. 444-448.

BACKLUND, A. V.
> Zusammenstellung einer Theorie der klassischen Dynamik
> und der neuen Gravitationstheorie von Einstein. *Arkiv för
> mathematik, astronomie, och fysik*, Stockholm, 1919, Band
> 14, no. 11, 64 seite.

BATEMAN, H.
> General Relativity Theory. *Phil. Mag.*, Feb., 1909, vol.
> 37.
> Applies the theory to life and mind.

BLANCK, MAX.
> Zur Dynamik bewegter Systeme. Königlich Preussischen
> Akademie der Wissenschaften. *Sitzungbericht, Berlin*, 1907,
> pp. 542-570.
> His theory.

BRILLOUIN, MARCEL.
> Propos Sceptiques au Sujet du Principe de **Relativité**
> *Scientia*, Bologna, 1913, vol. 13, pp. 10-26.

BROSE, HENRY L.
> Einstein's Theory of Relativity (non-math. form). Lec-
> ture published in pamphlet by B. H. Blackwell. Qxt. Not
> in N. Y. Public. Noted in *English Mechanic*, Dec. 19,
> 1919.

CARMICHAEL, ROBERT DANIEL.
> The Theory of Relativity. *Mathematical Monographs*.
> no. 12. New York : John Wiley, 1913.

COBB, Charles W.
> Relativity. *Journal of Philosphy, Psycholgy, and Scien-
> tific Method*, Jan 18, 1917.

CONWAY, A. W.
> *Relativity*. *Edin. Math. Tracts*, no. 3. London. 1913.

CROMMELIN, A. C. D.
> Results of the Total Solar Eclipse of May 29 and **the**
> Relativity Theory. *Science*, vol. 50, pp. 518-520. **Sci.**
> *Amer. Supp.*, Dec 6, 1919.

CUNNINGHAM, E.
  Report on the Relativity Theory of Gravitation. London
  Green, 1915.
DeZUANI, ARMANDO.
  Equilibrio relativo ed equazioi gravitazionali di Einstein
  nel caso stazionario. *Il nuovo cimento*, Pisa, 1919, vol. 18,
  pp. 5-25, July, 1919.
DONDER, T. DE.
  Gravitational Tensors of Einstein's Theory. *Science Ab-
  stracts*, April 30, 1919.
DROSTE, J.
  The Field of Moving Centres in Einstein's Theory of
  Gravitation. Koninklijke Akadamie van Wetenschappen.
  Proceedings, Amsterdam, 1916, vol. 19, pp. 447-475.
  The field of a single Centre in Einstein's Theory of
  Gravitation and the Motion of a Particle in that Field.
  Koninklijke Akademie van Wetenschappen. *Proceedings*,
  Ansterdam, 1916, vol. 19, pp. 197-218.
EDDINGTON, A. S.
  Report on the Relativity Theory of Gravitation. London :
  Fleetway Press, 1918. 91 p.
  The only account of Einstein's latest theory in English.
EDDINGTON, A. S., and OTHERS.
  The Deflection of Light by Gravitation and the Einstein
  Theory of Relativity. The Report of the British Eclipse
  Expedition to the Royal Society. *Scientific Monthly*, Jan.,
  1920.
GROSSMAN, MARCEL.
  Définitions, méthodes et problèmes mathématiques relatifs
  à la théorie de la gravitation. *Archives des Sciences
  Physiques et Naturelles*, Genève, Jan.-Juin, 1914, 4 période,
  Tome 37, pp. 13-19.
GUILLAUME, E.
  Theory of Relativity. *Archives des Sciences*, Dec., 1918,
  vol. 46.
HUMM, R. J.
  Energy Equations of General Theory of Relativity. *An-
  nalen der Physik*, 1919, vol. 58.
HUNTINGTON, EDWARD V.
  A new approach to the theory of relativity. Festschrift
  Heinrich Weber. Leipzig, 1912, pp. 147-169.
ISHIWARA, JUN.
  Zür relativistischen Theorie der Gravitation. Tohoku Im-
  perial University. *Science* reports, ser. 1, Sendai, Japan,
  vol. 4. pp. 111-160.
KAFKA, H.

Tensor Analysis. *Annalen der Physik*, 1919, vol. 58, pp. 1-54.

KOTTLER, FRIEDRICH.
Uber die physikalischen Grundlagen der einsteinschen Gravitätionstheorie. *Annalen der Physik*, Leipzig, 1918, Band 56, Folge 4, pp. 408-462.

LARMOR, J.
Essence of Physical Relativity. *Nat. Acad. Sci. Proc.*, vol. 4, Nov., 1918.

LAUE, RD. M.
Das Relativitätsprinzip. Braunschweig, 1911. 208 p.

LECORNU, LEON.
La Mécanique. Paris: Flammarion, 1909.

LORENTZ, H. A.
Considerations élémentaires sur le principe de relativité *Revue générale des sciences*, Paris, 1914. Tome 25, pp. 179-186.
La gravitation. *Scientia*. Bologna, 1914, vol. 6, pp. 28-59.

LORENTZ, H. A.
Das Relativitätsprinzip. 1914.

LORENTZ, H. A.
La Gravitation. *Scientia*, Bologna, 1914, vol. 16, pp. 28-59.

LORENTZ, H. A.
On Einstein's theory of gravitation. Koninklijke Akademie van Wetenschappen. *Proceedings*. Amsterdam, 1917-18, vol. 19. pp. 1341-1369; vol, 20, p. 2-34.
On Hamilton's principle in Einstein's theory of gravitation. Koninklijke Akademie van Wetenschappen. *Proceedings*, Amsterdam, 1917, vol. 19. pp. 751-765.

LUNN, ARTHUR C.
Some functional equations in the theory of relativity. American Mathematical Society. *Bulletin*, Lancaster, Pa., 1919, vol. 26, pp. 26-34.

MICHELSON, A. A., and MORLEY, E. W.
The Relative Motion of the Earth and the Luminiferous Ether. *American Journal of Science*, 1887, vol. 34, p. 333. See also: *Astro-Physical Journal*, vol. 37, pp. 100-193. The Report of the crucial experiments that upset the idea of a stationary ether.

MINKOWSKI, HERMANN.
Relativitätsprinzip. *Jahresbericht der Deutschen Mathematiker-Vereinigung*. Leipzig, 1915, Band, 24, pp. 1241-1244.

NORDSTROM GUNNAR.
Einstein's theory of gravitation and Hergloto's mechanics of continua. Koninklijke Akademie van Wetenschappen

Proceedings, Amsterdam, 1917. vol. 19.

NORDSTROM, G.
>On the Energy of the Gravitation Field in Einstein's Theory. Koninklijke Akadamic van Wettenschappen. *Proceedings*, Amsterdam, 1918, vol. 20. pp. 1238-1245.

PALATINI A.
>La Teoria di Relativita nel suo Sviluppo Storico. *Scientia*, Nov. and Dec. 1919, (in Italian and French).

PETZOLDT, J.
>Does the Relativity Theory Exclude the Reality of Space and Time? *Deutsche Phys. Gesell.*, Dec. 30, 1918.

RITZ, W.
>La gravitation. *Scientia*, Bologna, 1909, vol. 5, pp. 152-165.

RUFUS, CARL.
>Relativity in astronomy. *Popular Astronomy*, Northfield, Minn., 1918, vol. 26, pp. 160-165.

SILBERSTEIN, L.
>The Theory of Relativity. London: Macmillan, 1914. 92 p.

SITTER, W. DE.
>On Einstein's theory of gravitation, and its astronomical consequences. Royal Astronomical Society. *Monthly Notices*, London, 1915-18, vol. 76, pp. 699-728 : vol. 77, pp. 155-184; vol. 78, pp. 3-28.

>On the relativity of rotation in Einstein's theory. Koninklijke Akademie van Wetenschappen. *Proceedings*, Amsterdam, 1917, vol. 19, pp. 527-532.

>On the relativity of inertia. Remarks concerning Einstein's latest hypothesis. Koninlijke Akademie van Wetenschappen. *Proceedings*, Amsterdam, 1916, vol. 19. pp. 367-381.

>Planetary motion and the motion of the moon according to Einstein's theory. Koninklijke Akademie van Wetenschappen. *Proceedings*, Amsterdam, 1916, vol. 19. pp. 307-381.

>Further Remarks on the Solutions of the Field-equations of Einstein's Theory of Gravitation. Koninklijke Akademie van Wetenschappen. *Proceedings*, Amsterdam, 1918, vol. 20. pp. 1309-1312.

ST. JOHN, CHARLES EDWARD.
>The principle of generalized relativity and the displacement of the Fraunhofer lines toward the red. Carnegie Institution of Washington. *Contributions of the Mount Wilson Solar Observatory*, Washington, vol 7, no. 138.

TRESLIN J.
>The equations of the theory of electrons in a gravitation

field of Einstein deduced from a variation principle. Koninklijke Akademie van Wetenschappen. *Proceedings.* Amsterdam, 1917, Band 19, pp. 892-896.

TUNZELMAN, G. W. DE.
  Relativity Theory. *Science Progress,* 1919, vol. 13.

WEYL, H.
  Eine neue Erweiterung der Relativitätstheorie. *Annalen der Physik,* 1919, vol. 59, pp. 100-123.
  Raum, Zeit, Materie. Springer. 1918.
  Zur Gravitationstheorie. *Annalen der Physik,* 1917, vol. 54, pp. 117-145.

WILSON, EDWIN BIDWELL.
  Generalized coördinates, relativity, and gravitation. *Astrophysical Journal,* Chicago, 1917, vol. 45, pp. 244-258.

WILSON, WILLIAM.
  Relativity and gravitation. Physical Society of London. *Proceedings,* London, 1919, vol. 31, pp. 69-77.

WILSON and LEWIS.
  The Time-Space Manifold. *Proceedings of the American Academy of Arts and Sciences,* Nov., 1912, vol. 48, pp. 389-506.

Most of the mathematical references cited above have been borrowed bodily from the book list prepared by Miss Mary E. Todd of the Science Room of the New York Public Library, and published in the *Library Journal.* As soon as the Einstein craze struck New York these books were placed on a long table and it has been difficult to find a seat at this table, day or evening, ever since. At Cambridge University when Professor Eddington lectured on the Einsten theory the students waiting for the opening of the hall doors formed a cue extending half-way across Trinity Great Court. It is unusual in any university to have " standing room only " at a lecture on mathematical physics.

About half of the present volume appeared in *The Independent* of November 29, December 7, 13, and 20, 1919, and I am indebted to Hamilton Holt, the editor, and to Karl V. S. Howland, the publisher of that magazine, for the privilege of reprinting them in book form. I am further grateful to several professors of physics, mathematics, astronomy and philosophy who have been kind enough to criticize and correct this material, but since it would not be fair to hold them responsible for my personal views and unconventional language I shall have to express my thanks to them in private.

W. JOLLY AND SONS, LTD., PRINTERS, ABERDEEN.